AFC
CABLE SYSTEMS

Introduction

This installation manual is intended to provide a basic, yet, comprehensive overview of the major considerations related to Code-complying installations and cost-effective applications of armored cable (Type AC) and interlocked-armor metal-clad cable (Type MC) as feeders and branch-circuits to supply lighting, motors, appliances, machines, electric heating, fans, air-conditioners, and refrigeration equipment, as well as for signaling and control. The excellent safety record of both Type AC and Type MC cable, as well as their durability and reliability make armored and metal-clad cable the preferred wiring methods of many design and consulting electrical engineers, contractors, and plant electrical personnel for a wide variety of commercial, industrial, institutional, and residential applications.

Overview

Chapter One 1

Chapter 1 provides general requirements for selecting the right cable type to assure adequate capacity and optimum layout of feeders and branch-circuits. Topics covered include how to ground and bond equipment, the differences and similarities between AC and MC cable, as well as permitted and prohibited use for both types of cables.

Chapter Two 19

Chapter 2 discusses in detail how to determine conductor ampacity, based upon condition of use. Derating techniques for more than three current carrying conductors and for surrounding ambient temperatures exceeding 86°F are shown with step by step procedures for finding allowable ampacities.

Chapter Three 29

Chapter 3 focuses on the correct methods involved in selecting the proper size overcurrent protection device, based upon the terminals and allowable ampacities of the conductors in each cable run.

Chapter Four **41**

Chapter 4 covers the concerns that are more directly associated with the actual installation of AC and MC cable. Requirements, such as layout of boxes, routing of and supports of cables, along with protection schemes, are to ensure cables are not damaged.

Chapter Five **65**

Chapter 5 highlights the wiring of special equipment and circuits such as alarm systems, security systems, communication systems, and other special types of equipment pertinent to industry use.

Chapter Six **79**

Chapter 6 sets forth the rules and regulations for armored cable, utilized in special facilities and locations.

Chapter Seven **99**

Chapter 7 is concerned with the benefits provided by armored cable for the remodeling or installation of new circuits in existing premises.

Summary

This manual presents concepts and techniques based on the requirements of the National Electrical Code® (NEC®) and Underwriters Laboratories (UL) requirements, as well as modern day design and installation practices, which are greatly influenced by NEC® rules.

Each chapter contains illustrations validated with Sections of the NEC® to provide fast and easy references to the rules that govern the use of AC and MC cable.

To achieve the maximum benefit from this manual, it is recommended to read through it first. It is best to have an overall picture of how to properly and effectively install AC and MC cable before trying to comprehend the rules and regulations of the NEC® and UL.

By reading and studying this manual and the Sections referred to in the NEC®, the user of this manual will have a better understanding of how armored cable is to be used and installed.

GENERAL REQUIREMENTS

This chapter deals with the construction of armored cables which includes types available, types of conductors, grounding techniques, and the locations where such cable can or cannot be installed.

CABLE CONSTRUCTION AND USE

Although armored cable and interlocked-armor metal-clad cable have been in use for a number of years, many electrical professionals cannot readily distinguish one from the other. This is due to the fact that the overall spiral-wound metal sheath gives a similar outward appearance. However, that is about where the similarities end. There are very definite differences in construction. The uses permitted for each vary and it is critical that designers, installers, and inspectors be able to identify a given metal-sheathed cable assembly as either Type AC or Type MC.

TYPE AC CABLE

The NEC defines AC cable as a fabricated assembly of insulated conductors in a flexible metallic enclosure.

The safety, durability, and reliability of armored cable represent only part of the benefits provided by the selection of AC cable over raceways. In addition to excellent performance characteristics, AC cable provides substantial labor savings because it is flexible, easy to handle, and can be installed quickly. The pre-wired cable assemblies eliminate the need to pull conductors into a raceway, which in turn greatly reduces the possibility of conductor damage during pulling. By selecting AC cable to wire modern electrical systems, both labor and material savings are achieved.

The make-up of AC cable consists of an overall metal sheath with the following:

1. Insulating bushing (anti-short).
2. 90°C THHN insulated, copper conductors.
3. A moisture-resistant fibrous covering around the conductors.
4. No. 16 AWG aluminum bonding wire.
5. Overall interlocked metal armor.

See Figure 1-1 for a detailed description of the way AC cable is built, per NEC 333-19 and NEC 333-20.

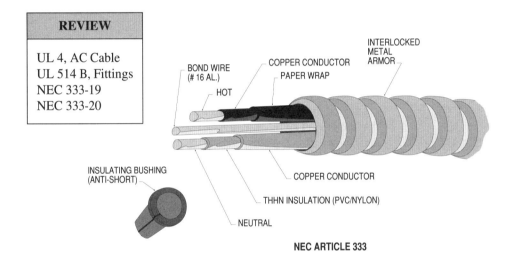

REVIEW
UL 4, AC Cable UL 514 B, Fittings NEC 333-19 NEC 333-20

Figure 1-1. AC cable shall be listed and comply with both NEC and UL requirements.

CONDUCTORS

In the UL White Book (General Information for Electrical Construction) under armored cable (AC cable), it states that the insulation of conductors may be either a thermoplastic or thermoset type insulation.

The insulated conductors of Type AC cable shall be of a type listed in Table 310-13 or those identified for use in this cable. AC cable conductors shall have an overall moisture-resistant and fire-retardant fibrous covering. Most commercially available AC cable is provided with a thermoplastic insulation (usually 90°C rated THHN) and most AC cable is listed and labeled as type ACTHH.

Note: AC cable is limited to no more than four insulated conductors plus a grounding wire. (Per UL Std. 4)

ACTHH cable requires individual conductors to have a moisture-resistant fibrous covering. When determining the ampacity of AC cable per NEC 310-15(a), the use of Ampacity Tables 310-16 thru 310-19 and their accompanying rules for derating must be utilized.

The size conductors in AC cable are No. 14 AWG through No. 1 AWG copper.

See Figure 1-2 for a detailed description of the types of insulations and use of conductors in AC cables per NEC 333-20, Tables 310-13 and 310-16.

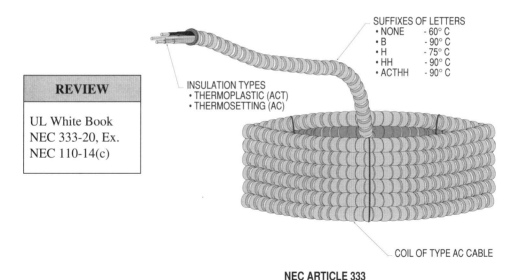

SUFFIXES OF LETTERS
- NONE   - 60° C
- B   - 90° C
- H   - 75° C
- HH   - 90° C
- ACTHH   - 90° C

INSULATION TYPES
- THERMOPLASTIC (ACT)
- THERMOSETTING (AC)

COIL OF TYPE AC CABLE

NEC ARTICLE 333

REVIEW
UL White Book NEC 333-20, Ex. NEC 110-14(c)

Figure 1-2. AC cable may employ different types of insulations with 60°, 75°, or 90°C ratings, but today almost all is 90° rated. *Note: The ampacity of AC cable is taken at 60°C, per NEC 333-20, Ex. and Table 310-16, when installed in thermal insulation.*

GROUNDING

The most distinctive physical characteristic of AC cable is the No. 16 AWG aluminum bond wire in conjunction with the metal armor which serves to provide the low-impedance fault-return path required to facilitate the operation of overcurrent protection devices under ground-fault conditions. The bond wire can be copper, per the NEC. However, UL's General Information Directory requires the use of aluminum for the bond wire. NEC 250-91(b)(6) allows the armor of AC cable to serve as the equipment grounding means to ground metal boxes, enclosures, etc. of the electrical system to a single grounding point. **(See Note)**

Note: Armored cable will always have an aluminum bond wire under the armor and the individually insulated conductors will each have a Kraft paper wrap.

The armor and bond wire combination with listed fittings for grounding works together to clear a ground-fault and is recognized as an equipment grounding means in accordance with UL 4, for AC cable.

See Figure 1-3 for a detailed description of how this bonding strip is used to aid in clearing a ground-fault per NEC 333-19 and 333-21.

REVIEW
NEC 300-3(b)
NEC 300-5(i)
NEC 300-20
NEC 333-21
NEC 250-91(b)
NEC 250-57(b)

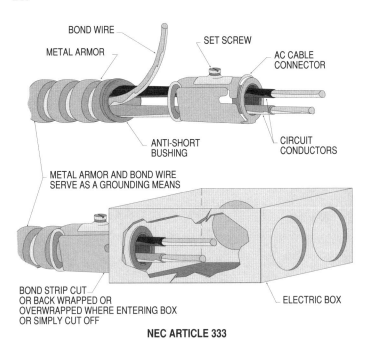

BOND WIRE
METAL ARMOR
SET SCREW
AC CABLE CONNECTOR
ANTI-SHORT BUSHING
CIRCUIT CONDUCTORS
METAL ARMOR AND BOND WIRE SERVE AS A GROUNDING MEANS
BOND STRIP CUT OR BACK WRAPPED OR OVERWRAPPED WHERE ENTERING BOX OR SIMPLY CUT OFF
ELECTRIC BOX

NEC ARTICLE 333

Figure 1-3. The armor and bond wire combination is recognized as an acceptable equipment grounding means.

NEC 250-51 requires AC cable to provide a path to ground from circuits, equipment, and metal enclosures for conductors and shall be:

1. permanent and continuous,
2. have capacity to conduct any fault current likely imposed on it, and
3. have sufficiently low impedance to limit the voltage to ground and to facilitate the operation of the circuit protective devices.

When installing AC cable with a metallic box, the devices shall be pigtailed from the grounding screw-terminal of the device to a tapped grounding screw provided within the box. When installing self-grounded devices, the device is grounded through its connection to the box.

Note: Tables 4.1 and 4.2 in UL 4 list the smallest equipment grounding conductors that are acceptable. Sometimes the equipment grounding conductor is larger than required by the NEC.

Equipment grounding conductors (when installed) in AC cables are used to ground the noncurrent-carrying metal parts of equipment, and are added to help keep the equipment at zero potential, and provide a redundant path for the ground-fault current. Equipment grounding conductors also ensure the safety of personnel from electrical shock. The size of the equipment grounding conductor is based on the size of the overcurrent protection device protecting the circuit conductors.

For example: If a 30 amp circuit breaker or set of fuses protect the conductors supplying a branch-circuit, Table 250-95 is utilized to size the equipment grounding conductor. Referring to Table 250-95, a 30 amp overcurrent protection device requires a #10 copper, equipment grounding conductor. See armored cables in UL 4. **(See Note)**

MARKING

The NEC covers markings that the manufacturer is required to provide. In addition to a tag on the coil or reel indicating the cable's voltage rating, insulation type, AWG-size, and the manufacturer's name, the NEC requires an external marking on the cable armor, throughout its entire length, that will provide for "ready identification of the manufacturer".

The insulating or "anti-short" bushing is not on the cable when it is delivered. Such bushings are usually provided in bags attached to the coil or reel and are installed at the time of termination to protect the conductors "at all points where the armor of AC cable terminates".

See Figure 1-4 for a detailed description of how AC cable terminations are made at boxes with fittings per NEC 333-9 and NEC 333-22 for marking requirements.

5

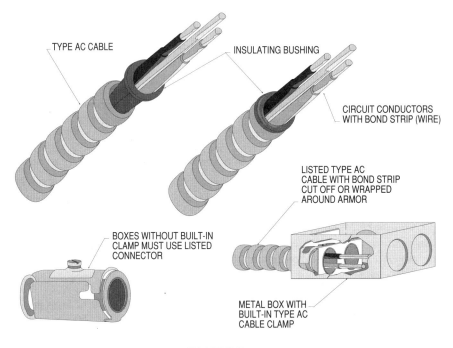

TYPE AC CABLE

INSULATING BUSHING

CIRCUIT CONDUCTORS
WITH BOND STRIP (WIRE)

LISTED TYPE AC
CABLE WITH BOND STRIP
CUT OFF OR WRAPPED
AROUND ARMOR

BOXES WITHOUT BUILT-IN
CLAMP MUST USE LISTED
CONNECTOR

METAL BOX WITH
BUILT-IN TYPE AC
CABLE CLAMP

NEC ARTICLE 333

REVIEW
NEC 300-15(c)
NEC 333-19
NEC 333-9
NEC 333-22
UL 514 B

Figure 1-4. The bond wire may simply be cut off at the end of the armor or wrapped around the armor and secured to the connector or pulled back over armor or terminated in box.

USES PERMITTED

AC cable shall be permitted in both exposed and concealed work for virtually all types of electrical systems and for branch-circuits and feeders, where not subject to physical damage.

AC cable shall be permitted in dry locations and embedded in plaster finish on brick or other masonry. AC cable installed in walls that are not exposed or subject to excessive moisture or dampness or are below grade shall be permitted to be installed in the air voids of masonry block or tile walls.

AC cable may also be installed in environmental air handling spaces such as ceilings per NEC 300-22(c).

See Figure 1-5 for a detailed description of the permitted uses of AC cable per NEC 333-3.

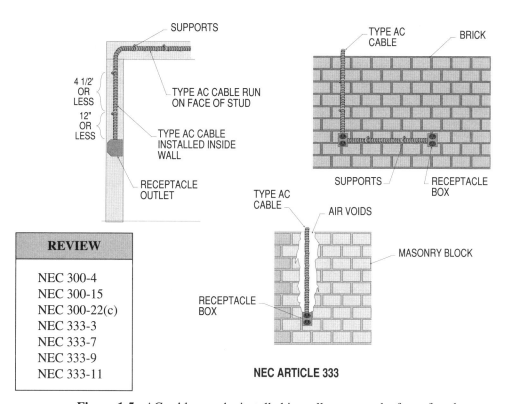

REVIEW
NEC 300-4
NEC 300-15
NEC 300-22(c)
NEC 333-3
NEC 333-7
NEC 333-9
NEC 333-11

NEC ARTICLE 333

Figure 1-5. AC cable may be installed in walls, run on the face of studs, embedded in plaster finish or brick, in dry air voids of tiles, etc.

USES NOT PERMITTED

Note: AC cable shall not be installed where embedded in plaster finish, or attached to brick or other masonry where such areas are considered to be a damp or wet location. See NEC 333-4 for installing AC cable in non-permitted locations.

AC cable shall not be installed in wet or damp locations. Other locations where AC cable shall not be installed are as follows:

1. In theaters and similar locations,
2. Places of assembly,
3. Motion picture studios,
4. Hazard (classified) locations,
5. Exposed to corrosive fumes or vapors,
6. Cranes or hoists,
7. Storage battery rooms,
8. In hoistways or on elevators, and
9. Commercial garages.
    * Some exceptions are permitted in NEC Section 333-4.

TYPE MC CABLE

The NEC defines MC cable as a factory assembly of one or more conductors, each individually insulated and enclosed in a metallic sheath of interlocking tape or a smooth or corrugated tube.

The make-up of a typical interlocked armor MC cable consists of the following:

1. An overall metal sheath,
2. An overall cable tape,
3. A marker tape,
4. THHN insulation,
5. Copper circuit conductors, and
6. Equipment grounding conductors.

The basic differences between MC and AC cable is that MC cable does not have an aluminum bond wire under the armor and the individually insulated conductors do not have a Kraft paper wrap.

The conductors may be copper, aluminum, or copper-clad aluminum. They may be solid or stranded in sizes up to No. 8. Conductors must be stranded in No. 6 and larger. Conductors shall not be smaller than No. 18 for copper, and No. 12 for aluminum or copper-clad aluminum. AFC cables are made with copper conductors.

The metallic sheath of MC cable may take different forms. For example, it may be a smooth tube, a corrugated tube, or an interlocked-spiral wound armored tape.

All interlocked-armor MC cables are required to have an equipment grounding conductor that may be bare or insulated and may be sectioned. The required equipment grounding conductor in conjunction with the cable armor and connector serve to satisfy the rule for a low-impedance fault-return path to clear the overcurrent protection device due to a ground-fault. This manual addresses the interlocked armor type.

See Figure 1-6 for a detailed description of the way MC cable is built, per NEC 334-20 thru NEC 334-22.

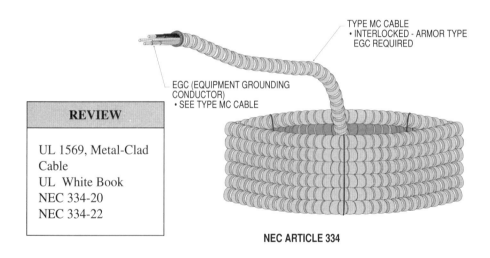

TYPE MC CABLE
• INTERLOCKED - ARMOR TYPE
EGC REQUIRED

EGC (EQUIPMENT GROUNDING CONDUCTOR)
• SEE TYPE MC CABLE

NEC ARTICLE 334

REVIEW
UL 1569, Metal-Clad Cable UL White Book NEC 334-20 NEC 334-22

Figure 1-6. MC cable is available with different types of metal armor and equipped with a variety of conductors to suit modern day wiring needs.

CONDUCTORS

The UL White Book, under metal-clad cable (MC cable), indicates that the insulation of conductors be either thermoplastic or thermoset.

MC cable insulated conductors shall be of a type listed in 310-13 or those identified for use in this cable. MC cable conductors shall have an overall moisture-resistant and fire-retardant cable tape. Most commercially available MC cable is provided with a thermoplastic insulation (usually 90°C rated THHN).

When determining the ampacity of MC cable, per NEC 310-15(a), the use of ampacity Tables 310-16 thru 310-19 and their accompanying requirements for derating are utilized per Note 8(a).

See Figure 1-7 for a detailed description of the way Type MC cable is built per NEC 334-20, Table 310-13 and Table 310-16.

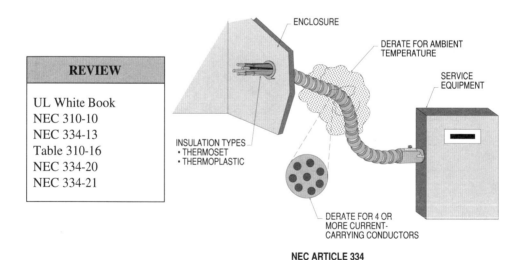

REVIEW

UL White Book
NEC 310-10
NEC 334-13
Table 310-16
NEC 334-20
NEC 334-21

ENCLOSURE

DERATE FOR AMBIENT TEMPERATURE

SERVICE EQUIPMENT

INSULATION TYPES
• THERMOSET
• THERMOPLASTIC

DERATE FOR 4 OR MORE CURRENT-CARRYING CONDUCTORS

NEC ARTICLE 334

Figure 1-7. The ampacity of the four conductors in the above MC cable is determined by the type of insulation, number of conductors and ambient temperature.

GROUNDING

MC cable can be used to ground the metallic enclosures and equipment which is accomplished through the proper installation of a listed grounding type connector and connection of the cable's equipment grounding conductor to the enclosure of equipment.

The metallic sheath of interlocked armor type MC cable in combination with a listed connector and the cable's equipment grounding conductor is recognized as a ground return path. As a result, attachment of the cable sheath by a listed connector and connection of the equipment grounding conductor to the enclosure are both required to provide proper grounding.

Where a device is supplied by an MC cable, the device's green hex-head (grounding terminal) must be connected to the box's grounding screw, or a pigtail can be connected to the box and both the device and pigtail can be connected by a wire to the cable's equipment grounding conductor. For self-grounding devices, it is important to remember that the cable's grounding conductor must ground the box. The device is then grounded through its connection to the box.

See Figure 1-8 for a detailed description of how to ground equipment using MC cable.

REVIEW
UL White Book NEC 334-22 NEC 334-23 NEC 250-91(b)

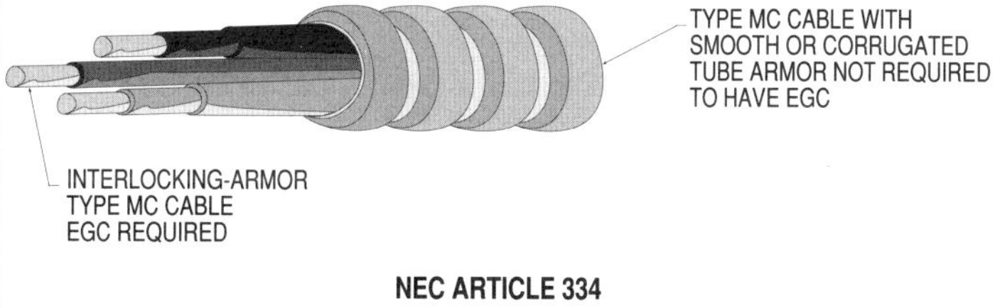

TYPE MC CABLE WITH
SMOOTH OR CORRUGATED
TUBE ARMOR NOT REQUIRED
TO HAVE EGC

INTERLOCKING-ARMOR
TYPE MC CABLE
EGC REQUIRED

NEC ARTICLE 334

Figure 1-8. MC cable of the interlocking armor type requires an equipment grounding conductor (EGC) for grounding.

11

MARKING

The NEC and UL cover markings that the manufacturer is required to provide. In addition to a tag on the coil or reel indicating the cable's voltage rating, insulation type, AWG-size, and the manufacturer's name, the NEC requires an external marking on the cable armor, throughout its entire length or an internal marker tape that will provide for " ready identification of the manufacturer".

The insulating or "anti-short" bushings are not required by NEC Article 334 but such bushings are usually provided in bags attached to the coil or reel and are installed at the time of termination to protect the conductors "at all points where the armor of MC cable terminates". These bushings may not be needed when connectors with built in insuliners are used.

See Figure 1-9 for a detailed description of the way MC cable is marked and how terminations are made at boxes with fittings per NEC 334-12 and NEC 334-24.

REVIEW
UL 1569, MC Cable
UL 514 B, MC Cable
NEC 334-12
NEC 334-24
NEC 300-15(c)

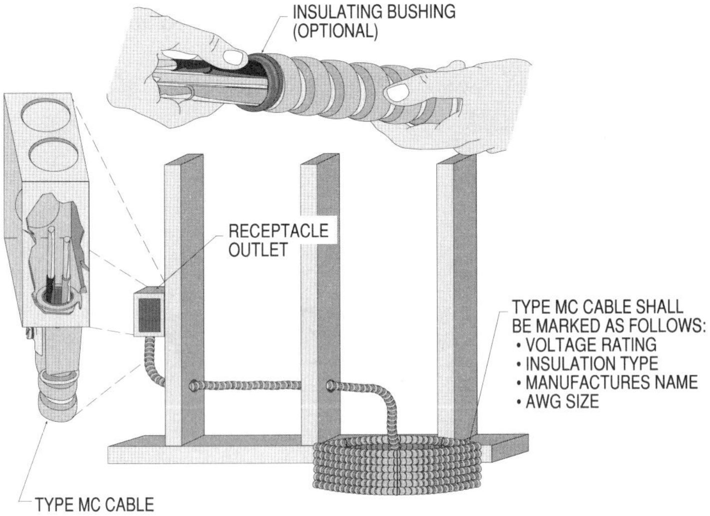

INSULATING BUSHING (OPTIONAL)

RECEPTACLE OUTLET

TYPE MC CABLE SHALL BE MARKED AS FOLLOWS:
• VOLTAGE RATING
• INSULATION TYPE
• MANUFACTURES NAME
• AWG SIZE

TYPE MC CABLE

NEC ARTICLE 334

Figure 1-9. MC cable of the interlocked armor type uses the combination of an equipment grounding conductor and the armor as the equipment grounding means according to NEC 250-91(b).

USES PERMITTED

MC cable shall be permitted in both exposed and concealed work for virtually all type of electrical systems and for branch circuits and feeders, where not subject to physical damage.

MC cable shall be permitted in dry locations. MC cable installed in walls that are not exposed or subject to excessive moisture or dampness or are below grade shall be permitted to be installed in the air voids of masonry block or tile walls.

(*) - Supplemental protection in the form of a jacket is required.

MC cable can be used as follows:

1. Services, feeders and branch-circuits.
2. Power, lighting, control, and signal circuits.
3. Indoors and outdoors.
4. Where exposed* or concealed.
5. Direct burial*.
6. In cable trays.
7. In any approved raceway.
8. As open runs of cable.
9. As aerial cable on a messenger cable*.
10. In hazardous locations as permitted in Articles 501, 502, 503, and 504 of the NEC.
11. In dry locations.
12. In wet locations when any of the following conditions are met:
    (a) The metallic covering is impervious to moisture.
    (b) A lead sheath or moisture impervious jacket is provided under the metal covering.
    (c) The insulated conductors under the metallic covering are approved for use in wet locations.
13. Environmental air handling ceilings (300-22(c)).

Note: The conditions which must be met in (12) above, must also apply to (5) above.

See NEC 334-3 for the rules pertaining to the installation of Type MC cable in different locations.

USES NOT PERMITTED

MC cable shall not be allowed to be installed and used in the following locations:

Note: A large responsibility is left to the Authority Having Jurisdiction to judge if the material is suitable for the conditions mentioned.

1. Buried in concrete (unless jacketed)

2. Directly buried in corrosive soil (unless jacketed)

3. Where exposed to corrosive acids or caustic alkalis (unless jacketed)

4. NEC 334-4, Ex.: Where the metallic sheath is suitable for the conditions or is protected by material suitable for the conditions, per NEC 334-4, Ex. *(See Note)*

See Figure 1-10 for a detailed description of the non-permitted use of MC cable per NEC 334-4 and Ex.

REVIEW
NEC 300-6
NEC 334-4
UL 1569
UL 514 B

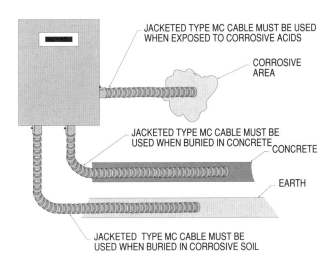

NEC ARTICLE 334

Figure 1-10. Jacketed MC cable must be used for locations such as soil, concrete, acids or caustic alkalis. This jacketed MC cable must be listed and labeled for the application. AFC manufactures such cable.

	Chapter One	
Answers	**Questions**	

<table>
<tr><td>T   F
a</td><td>1. The advantage of using AC cable over conduit is that there's savings of labor over raceways.</td></tr>
<tr><td>T   F</td><td>2. One of the purposes of the insulating bushing is to protect the conductors from physical damage.</td></tr>
<tr><td>T   F</td><td>3. The insulation for conductors in AC cable are usually THHN 90°C and their ampacity is selected at 60°C.</td></tr>
<tr><td>T   F</td><td>4. Type AC cable is limited to four or less conductors plus an equipment grounding conductor.</td></tr>
<tr><td>T   F</td><td>5. The metal armor of listed Type AC cable must not be used for grounding.</td></tr>
<tr><td>T   F</td><td>6. Each individual insulated conductor in Type AC cable has a kraft paper wrap.</td></tr>
<tr><td>T   F</td><td>7. The NEC requires Type AC cable to have a path to ground from circuits and metal enclosures.</td></tr>
<tr><td>T   F</td><td>8. When installing self-grounding devices the device is grounded through its connection to the box.</td></tr>
<tr><td>T   F</td><td>9. Type AC cable can be used in all places of public assemblies.</td></tr>
<tr><td>T   F</td><td>10. Type AC cable shall not be installed in the dry air voids in blocks or tiles.</td></tr>
<tr><td>_____</td><td>11. Type MC cable can have a metallic sheath of _____ tape or corrugated or smooth tube.</td></tr>
<tr><td>_____</td><td>12. Type MC cable usually has _____ cu. conductors for its circuits.</td></tr>
<tr><td>_____</td><td>13. Type MC cable with an interlocked-armor is required to have an _____ grounding conductor.</td></tr>
<tr><td>_____</td><td>14. Connectors used to terminate Type AC cable are required to be _____ for grounding.</td></tr>
</table>

Sections	Questions
_____	**15.** Bonding jumpers can be used to _____ the metal of a box to the yoke of the device.
_____	**16.** Type MC cable is _____ allowed to be buried in concrete.
_____	**17.** For Type MC cable installed in wet locations the metallic covering must be _____ to moisture.
_____	**18.** The insulation used with Type MC cable must be rated _____ .
_____	**19.** Type AC cable may be _____ in plaster finish or brick.
_____	**20.** Anti-short _____ are usually provided in bags attached to the coil or reel of AC cable.
_____	**21.** The smallest conductor usually permitted in Type AC cable is # _____ copper. **(a)** 22 **(b)** 18 **(c)** 16 **(d)** 14
_____	**22.** The largest conductor usually allowed in Type AC cable is # _____ copper. **(a)** 1 **(b)** 1/0 **(c)** 2/0 **(d)** 4/0
_____	**23.** The size of the bonding wire for Type AC cable is # _____ aluminum. **(a)** 22 **(c)** 16 **(b)** 18 **(d)** 14

	Chapter One
Sections	**Questions**
_____	**24.** The hex-head grounding terminal on the yoke containing a device is _____ in color. **(a)** Silver **(b)** Copper **(c)** Red **(d)** Green
_____	**25.** Most commercially available Type MC cable is provided with _____ insulation. **(a)** Rubber **(b)** Thermoplastic **(c)** All of the above **(d)** None of the above

CONDUCTOR AMPACITY

One of the most fundamental tasks associated with electrical work is determining the ampacity of the conductors. The NEC recognizes that the maximum current carrying capacity for every circuit conductor is based upon their condition of use.

The industry utilizes two methods in calculating the ampacity of conductors. The first method is described in NEC 310-15(a) and the second method is detailed in NEC 310-15(b).

Because the first method is the most widely used, this chapter discusses the procedures in detail. The allowable ampacity of conductors are determined by actual comprehensive calculations based upon installation and use of AC and MC cables.

ALLOWABLE AMPACITY OF CONDUCTORS

The NEC allows no more than three current-carrying conductors in a raceway, cable, or buried side by side in the earth without derating. The NEC states that the ampacities of conductors listed are limited to a surrounding ambient temperature of 86°F or less. AC and MC cables enclosing more than three current-carrying conductors and routed through an ambient temperature of over 86°F are subject to derating. Correction factors must be applied based upon the number of current-carrying conductors and surrounding ambient temperature.

THREE OR LESS CONDUCTORS IN A CABLE

When installing AC or MC cable where the ambient temperature is not over 30°C (86°F) and consisting of not more than three current-carrying conductors within the cable, the conductors ampere rating are sized in amperes by the appropriate ampacity Table based on the AWG size and insulation type.

Conductors routed in cables as mentioned above are considered proper and the ampacities in Table 310-16 may be used to supply electrical loads.

For example: A No. 12-2 AC cable with THHN copper conductors has an ampacity of 20 amps based upon a 20 amp overcurrent protection device per the obelisk to Table 310-16.

See Figure 2-1 for a step by step procedure for selecting the allowable ampacity of three or less conductors in a cable that are current-carrying and routed through an ambient temperature of 30°C (86°F) or less per Note 8(a) to ampacity Tables 0 -2000 volts.

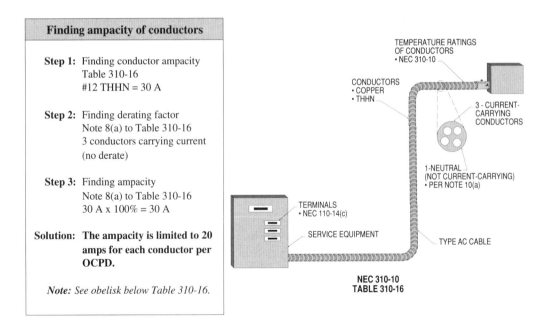

Finding ampacity of conductors

Step 1: Finding conductor ampacity
Table 310-16
#12 THHN = 30 A

Step 2: Finding derating factor
Note 8(a) to Table 310-16
3 conductors carrying current
(no derate)

Step 3: Finding ampacity
Note 8(a) to Table 310-16
30 A x 100% = 30 A

Solution: **The ampacity is limited to 20 amps for each conductor per OCPD.**

Note: See obelisk below Table 310-16.

TEMPERATURE RATINGS
OF CONDUCTORS
• NEC 310-10

CONDUCTORS
• COPPER
• THHN

3 - CURRENT-
CARRYING
CONDUCTORS

1-NEUTRAL
(NOT CURRENT-CARRYING)
• PER NOTE 10(a)

TERMINALS
• NEC 110-14(c)

SERVICE EQUIPMENT

TYPE AC CABLE

NEC 310-10
TABLE 310-16

Figure 2-1. There is no derating the ampacity of conductors listed in Table 310-16, if the cable has three or less current carrying conductors and runs through an ambient temperature of 86° F.

THREE OR LESS CONDUCTORS WITH HIGH AMBIENT TEMPERATURE

When installing AC cable, where the ambient temperature is higher than 30°C (86°F) and consisting of not more than three current carrying conductors within the cable, the conductors amperes are adjusted according to the ambient temperature "correction factors" given at the bottom of Table 310-16.

If the ambient temperature is lower than 30° (86°F), the ambient temperature "correction factors" are applied, based upon the size of the conductors and types of insulations.

For example: A cable containing three or less current carrying copper conductors with THHN insulation has a derating factor of 91%, when routed through an ambient temperature of 101° F. To obtain the correct allowable ampacity of the conductors, the ampacity in Table 310-16 for a No. 12 THHN copper conductor must be derated by 91%.

See Figure 2-2 for a step by step procedure for calculating the ampacities of conductors passing through ambient temperatures exceeding 30°C (86°F), with cables containing three or less current carrying conductors per correction factors below Table 310-16.

Finding ampacity of conductors

Step 1: Finding conductor ampacity
Table 310-16
#12 THHN = 30 A

Step 2: Finding correction factor
Table 310-16
101° F requires 91%

Step 3: Finding ampacity
Ampacity correction factors
Table 310-16
30 A x 91% = 27.3 A

Solution: **The ampacity is limited to 27.3 amps for each conductor.**

Note: See obelisk below Table 310-16 for 20 amp limitation.

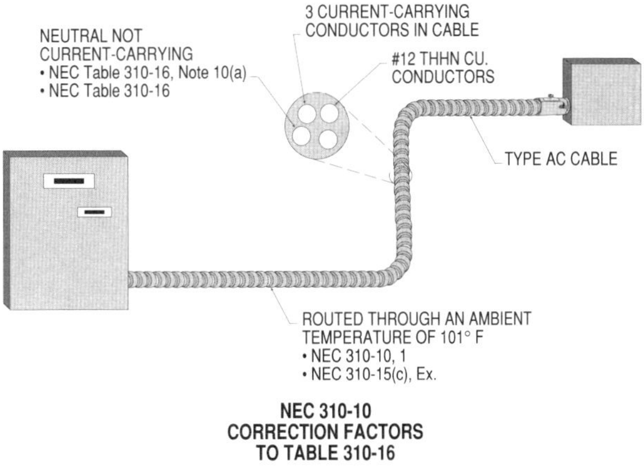

NEUTRAL NOT CURRENT-CARRYING
• NEC Table 310-16, Note 10(a)
• NEC Table 310-16

3 CURRENT-CARRYING CONDUCTORS IN CABLE
#12 THHN CU. CONDUCTORS

TYPE AC CABLE

ROUTED THROUGH AN AMBIENT TEMPERATURE OF 101° F
• NEC 310-10, 1
• NEC 310-15(c), Ex.

NEC 310-10
CORRECTION FACTORS
TO TABLE 310-16

Figure 2-2. Calculating the ampacity of conductors, where they are routed through ambient temperatures above 86° F, with only three current carrying conductors in the cable.

FOUR OR MORE CONDUCTORS IN A CABLE

Note: Designers recommend that no more than nine current-carrying conductors be used in a cable. This recommendation is based upon a 70% derating factor instead of a 50% derate, where there are over nine current-carrying conductors in a cable.

When installing AC or MC cable where the ambient temperature is not over 30°C (86°F) and consisting of more than three current-carrying conductors within the cable, the conductors are reduced by the actual number of current carrying conductors within the cable per Note 8(a) of the ampacity Tables.

For example: A cable consists of four No. 12 THHN copper current-carrying conductors which is routed through an ambient temperature of 86°F. From Note 8(a) to ampacity Tables 0-2000 volts, the ampacity per Table 310-16 must be derated to 80%, based upon 4 thru 6 current carrying conductors. (See **Note**)

See Figure 2-3 for a step by step procedure for calculating the ampacity of a cable enclosing seven to nine current carrying conductors routed through an area not more than 86°F per Note 8(a) to ampacity Table 0-2000 volts and correction factors below Table 310-16.

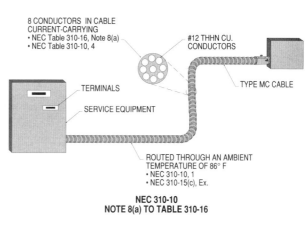

Finding ampacity of conductors

Step 1: Finding conductor ampacity
Table 310-16
#12 THHN = 30 A

Step 2: Finding derating factor
Note 8(a) to Table 310-16
8 conductors require 70%

Step 3: Finding ampacity
Table 310-16, Note 8(a)
30 A x 70% = 21 A

Solution: **The ampacity is limited to 21 amps for each conductor.**

Note: *See obelisk Table 310-16 for 20 amp limitation.*

8 CONDUCTORS IN CABLE
CURRENT-CARRYING
• NEC Table 310-16, Note 8(a)
• NEC Table 310-10, 4

#12 THHN CU.
CONDUCTORS

TERMINALS

TYPE MC CABLE

SERVICE EQUIPMENT

ROUTED THROUGH AN AMBIENT
TEMPERATURE OF 86° F
• NEC 310-10, 1
• NEC 310-15(c), Ex.

NEC 310-10
NOTE 8(a) TO TABLE 310-16

Figure 2-3. The ampacity listed in Table 310-16 must be derated per "correction factors" to Table 310-16, where a cable is routed through areas of high ambient temperatures.

DOUBLE DERATING CONDUCTORS IN A CABLE

When installing MC cable where the ambient temperature is higher than 30°C (86°F) and consisting of more than three current-carrying conductors within the cable, the conductor's ampacity must be adjusted according to the ambient temperature "correction factors" and reduced by the actual number of current carrying conductors within the cable per Note 8(a) of the ampacity Tables.

For example: A No. 12-8 w / ground MC cable is run through an ambient temperature of 120°F with all THHN copper conductors carrying current. From Note 8(a) to ampacity Tables 0-2000 volts, each conductor must be derated to 80% of its ampacity per Table 310-16. An additional derating of 76% must be applied per "correction factors" below Table 310-16, due to the surrounding ambient temperature of 125°F.

See Figure 2-4 for a step by step procedure for calculating the allowable ampacities of conductors based on four or more current carrying conductors exposed to surrounding ambient temperatures exceeding 30°C (86°F) per correction factors below Table 310-16.

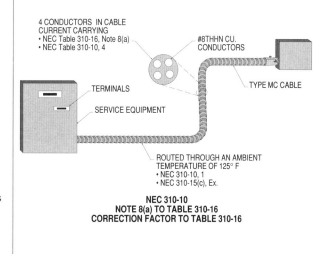

Finding ampacity of conductors

Step 1: Finding conductor ampacity
Table 310-16
#8 THHN = 55 A

Step 2: Finding derating factor
Note 8(a) to Table 310-16
4 conductors require 80%

Step 3: Finding ampacity
correction factors
Table 310-16
125° F requires 76%

Step 4: Finding ampacity
Note 8(a) & Correction factors
55 A x 80% x 76% = 33.44 A

Solution: **The ampacity is limited to 33.44 amps for each conductor.**

4 CONDUCTORS IN CABLE
CURRENT CARRYING
• NEC Table 310-16, Note 8(a)
• NEC Table 310-10, 4

#8THHN CU.
CONDUCTORS

TYPE MC CABLE

TERMINALS

SERVICE EQUIPMENT

ROUTED THROUGH AN AMBIENT
TEMPERATURE OF 125° F
• NEC 310-10, 1
• NEC 310-15(c), Ex.

NEC 310-10
NOTE 8(a) TO TABLE 310-16
CORRECTION FACTOR TO TABLE 310-16

Figure 2-4. Calculating the ampacity of conductors where they are routed through ambient temperatures above 86° F and the cable contains four or more current carrying conductors.

WHEN IS THE NEUTRAL CURRENT CARRYING

Note 10 to the ampacity Table 0-2000 volts addresses neutrals that are indeed true neutrals. Neutrals that carry only the unbalanced current for balanced single/phase, 3-wire or three/phase, 4- wire circuits supplying resistive loads are not required to be counted with the phase conductors for the purposes of applying Note 8(a) to Table 310-16. Neutral conductors used with three legs of a three/phase, 4- wire wye circuit that supply electronic discharge lighting, computers, peripherals, faxes, copiers, and similar electronic equipment must be counted as current carrying conductors. These neutrals are required to be counted under balanced conditions, because the neutral carries current that approximately equals, and in some cases exceeds, the phase current. This is a result of additive harmonic currents on the neutral. This type of load consists primarily of electronic equipment which uses the "Super Neutral" type MC cable which is available with neutrals to help compensate for this problem as follows:

Note: Neutrals may be striped to be identified with each phase conductor.

(1) A single oversized neutral, or
(2) A separate neutral for each phase conductor which is the same size as the phase conductor. **(See Note)**

Such neutrals count as current carrying conductors for the purpose of applying Note 8(a) to ampacity Tables 0-2000 volts.

Note 10(c) considers the neutral conductor to be current carrying where the major portion of the load consists of nonlinear loads such as electronic discharge lighting, computers, etc.

See Figure 2-5 for an example ot a cable supplying non-linear loads.

EQUIPMENT GROUNDING CONDUCTORS

Note 11 to the ampacity Tables 0-2000 volts indicates that equipment grounding conductors (EGC's) and bonding conductors are not counted when applying the derating rules of Note 8(a). The reasoning is the same as for true neutrals. Although these conductors occupy space, they do not normally carry current and therefore need not be counted for derating purposes.

The EGC's act as heat sinks and actually absorb heat from the other current carrying conductors.

NEHER-McGRATH METHOD

It is permissible to calculate ampacity in accordance with the Neher-McGrath method described in NEC 310-15(b) under engineering supervision.

As indicated by the definition of "ampacity" given in Article 100 of the NEC, the ampacity of a given conductor is the amount of current, in amperes, in which the conductor can carry continuously without exceeding its temperature rating.

The NEC approach to establishing the ampacity of any given conductor is aimed at designating the value of current that will cause the conductor to reach a thermal equilibrium and stabilize at a temperature no greater than the thermal limit of the insulation.

Software programs can be purchased to aid the designer wishing to utilize the Neher-McGrath concept of calculating the ampacity of conductors based upon their condition of use.

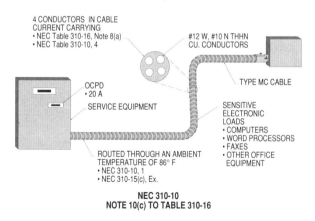

Figure 2-5. The neutral conductor is considered current carrying if the major portion of the load consists of harmonically related loads.

Chapter Two	
Answers	**Questions**
T F	**1.** Correction factors are applied based upon the number of conductors and surrounding ambient temperature.
T F	**2.** The NEC allows more than three current carrying conductors within Type AC cable when the ambient temperature is not over 30°C (86°F).
T F	**3.** If the ambient temperature is lower than 30°C (86°F), the ambient temperature "correction factors" are applied based upon the size of conductors and types of insulations.
T F	**4.** If the ambient temperature is higher than 30°C (86°F), the ambient temperature "correction factors" are given at the bottom of Table 310-16.
T F	**5.** Type AC cable consisting of more than three current carrying conductors within the cable, are reduced per Note 10(c) of the ampacity Tables.
T F	**6.** Where ambient temperatures are higher than 30°C (86°F) and consisting of more than three current carrying conductors within the cable, must be adjusted according to the ambient temperature "correction factors" and reduced by the actual number of current carrying conductors within the cable per Note 8(a) of the ampacity Tables.
T F	**7.** Note 10(c) requires the neutral conductor to be derated where the major portion (70%) of the load consists of nonlinear loads such as electric discharge lighting, computer, etc.
T F	**8.** Neutrals that are current carrying conductors shall count when applying Note 8(a) to ampacity Tables 0-2000 volts.
T F	**9.** Equipment grounding conductors and bonding conductors are required to be counted when applying the derating rules of Note 8(a).
T F	**10.** It is not permissible to calculate ampacity in accordance with Neher-M$^{\underline{c}}$Grath method.

	Chapter Two
Sections	**Questions**
_____	**11.** There is no derating the ampacity of conductors listed in Table 310-16, if the cable has _____ or less current-carrying conductors and runs through and ambient temperature of 86°F or less.
_____	**12.** When ambient temperatures are _____ than 30°C (86°F), the ambient temperature "correction factors" are applied with the given values at the bottom of Table 310-16.
_____	**13.** Type AC cable consisting of more than three current-carrying conductors within the cable, are derated per _____ of the ampacity Tables.
_____	**14.** Note 10(a) requires the neutral conductors to be derated where the major portion (_____%) of the load consists of nonlinear loads such as electric discharge lighting, computers, etc.
_____	**15.** Equipment grounding conductors and _____ conductors are not counted when applying the derating rules of Note 8(a).
_____	**16.** Type AC cable consisting of not more than three current-carrying conductors shall have an ambient temperature not over _____ . **a.** 84° **b.** 85° **c.** 86° **d.** 87°
_____	**17.** Type AC cable routed through an ambient temperature of 118°F shall have a percentage of _____% applied. **a.** 96% **b.** 82% **c.** 87% **d.** 91% **a.** 60%

18. Type AC cable consisting of four current-carrying conductors shall have a percentage of _____% applied.

a. 60%
b. 70%
c. 80%
d. 90%

19. What is the amperage for a No. 10-8 w/ground Type MC cable that is run through an ambient temperature of 120°F with all THHN copper conductors current-carrying.

a. 22.96 A
b. 23 A
c. 23.96 A
d. 24 A

20. What size neutral is required when supplying sensitive electronic equipment with a branch-circuit that has three No. 12 THHN copper current-carrying conductors.

a. 10 AWG
b. 14 AWG
c. 16 AWG
d. 8 AWG

CHAPTER 3

OVERCURRENT PROTECTION

After the allowable ampacity of a number of conductors in a cable have been calculated in accordance with the provisions of the NEC, a suitably rated overcurrent protective device (OCPD) must be selected to properly protect such conductors. The rating of overcurrent protective devices must comply with all the requirements in the NEC, pertaining to sizing and selecting such an OCPD.

SELECTING OCPD's

The basic rule of the NEC requires all conductors to be protected by a fuse or CB, with a rating equal to the conductors ampacity. However, calculated ampacity values do not always correspond directly to NEC standard OCPD ratings. It is permissible to use the next higher standard rated device. If this NEC rule is to be applied, the device must not be rated over 800 amps and the circuit must not be part of a multiwire branch-circuit supplying receptacle outlets used to cord-and-plug connect portable loads. For circuits rated over 800 amps, the fuse or circuit breaker must be equal to the conductors ampacity or the next lower standard rating must be selected.

Note: In cases where the OCPD does not provide overcurrent protection for the conductors in the cable, a second stage of protection for the conductors shall be provided such as overload relays, thermal protectors, etc. used to protect motor and A/C units.

See Figure 3-1 for a detailed description on selecting the proper OCPD based upon the conductor's allowable ampacity per NEC 240-3.

There are parts to NEC 240-3 (parts (a) through (c) and parts (d) through (m)) which address cases where a conductor does not have to be protected against overcurrent in accordance with its ampacity.

For example: Motor and air-conditioning branch-circuits and feeders may be protected as permitted in Articles 430 and 440 of the NEC. **(See Note)**

Note: When sizing a fuse or CB for No. 14, No. 12, and No. 10 copper conductors per the footnote (obelisk or dagger) to Table 310-16, the fuse or CB can never be selected at more than 15 , 20 , or 30 amps unless loads with in-rush currents are supplied, per NEC 240-3(e) thru (m).

See Figure 3-2 for a detailed description of sizing the OCPD to allow loads to start and run having high in-rush current per NEC 240-3. **(See Note)**

Remember, any derating factors that may be required, due to elevated ambients or more than three current-carrying conductors, or both must be applied against the allowable ampacity of Table 310-16 in determining the size OCPD's listed above.

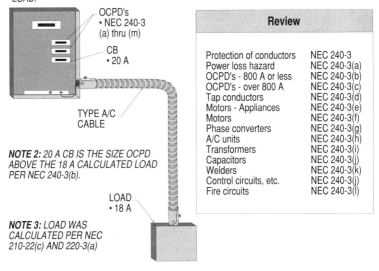

NOTE 1: 20 A IS SIZE CB ABOVE 18 A CALCULATED LOAD.

OCPD's
• NEC 240-3 (a) thru (m)

CB
• 20 A

TYPE A/C CABLE

NOTE 2: 20 A CB IS THE SIZE OCPD ABOVE THE 18 A CALCULATED LOAD PER NEC 240-3(b).

LOAD
• 18 A

NOTE 3: LOAD WAS CALCULATED PER NEC 210-22(c) AND 220-3(a)

Review	
Protection of conductors	NEC 240-3
Power loss hazard	NEC 240-3(a)
OCPD's - 800 A or less	NEC 240-3(b)
OCPD's - over 800 A	NEC 240-3(c)
Tap conductors	NEC 240-3(d)
Motors - Appliances	NEC 240-3(e)
Motors	NEC 240-3(f)
Phase converters	NEC 240-3(g)
A/C units	NEC 240-3(h)
Transformers	NEC 240-3(i)
Capacitors	NEC 240-3(j)
Welders	NEC 240-3(k)
Control circuits, etc.	NEC 240-3(j)
Fire circuits	NEC 240-3(l)

NEC 240-3(a thru l)

Figure 3-1. OCPD's may be greater in rating than the ampacity of the conductors in armored cables, if they supply loads having high in-rush currents to start and run motors or ampacity of conductors does not correspond to a standard OCPD.

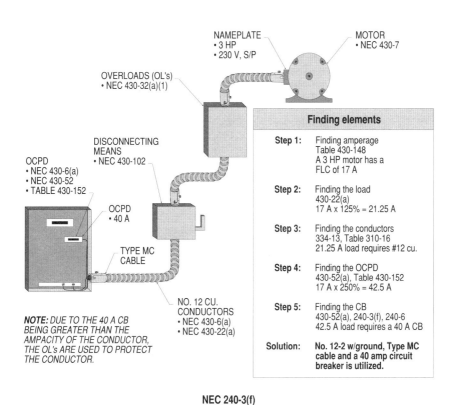

NAMEPLATE
• 3 HP
• 230 V, S/P

MOTOR
• NEC 430-7

OVERLOADS (OL's)
• NEC 430-32(a)(1)

DISCONNECTING
MEANS
• NEC 430-102

OCPD
• NEC 430-6(a)
• NEC 430-52
• TABLE 430-152

OCPD
• 40 A

TYPE MC
CABLE

NOTE: *DUE TO THE 40 A CB
BEING GREATER THAN THE
AMPACITY OF THE CONDUCTOR,
THE OL's ARE USED TO PROTECT
THE CONDUCTOR.*

NO. 12 CU.
CONDUCTORS
• NEC 430-6(a)
• NEC 430-22(a)

Finding elements	
Step 1:	Finding amperage Table 430-148 A 3 HP motor has a FLC of 17 A
Step 2:	Finding the load 430-22(a) 17 A x 125% = 21.25 A
Step 3:	Finding the conductors 334-13, Table 310-16 21.25 A load requires #12 cu.
Step 4:	Finding the OCPD 430-52(a), Table 430-152 17 A x 250% = 42.5 A
Step 5:	Finding the CB 430-52(a), 240-3(f), 240-6 42.5 A load requires a 40 A CB
Solution:	**No. 12-2 w/ground, Type MC cable and a 40 amp circuit breaker is utilized.**

NEC 240-3(f)

Figure 3-2. Because the OCPD is greater in rating than the conductors, a second stage of protection is required to protect the MC cable conductors and this protection is provided by the overloads in the control.

31

SIZING OCPD's

Note: Circuits capable of having loads added to the type armored cable system used must be protected by the OCPD at their allowable ampacities.

Another concern related to the operational characteristics of overcurrent protective devices is whether the load is continuously operating for 3 hours or more or noncontinuously operating for less than 3 hours. In selecting an overcurrent protective device to protect a conductor at a particular ampacity, the maximum permitted device rating and the conductor ampacity remain the same whether the circuit loading is continuous or noncontinuous. If the load is all noncontinuous, then the circuit may be loaded up to the rating of the overcurrent protective devices. However, if all or a portion of the load is continuous in nature, then the loading of the OCPD must be limited. In other words, the value of noncontinuous load plus 125% of the continuous load must not exceed the rating of the protective device, unless the device and assembly are listed and marked for use at 100% load for continuous operation.

For example: A continuous load of 15.9 amps must be increased up to 19.9 amps (15.9 A x 125% = 19.9 A) and the OCPD must be selected at 20 amps. **(See Note)**

See Figure 3-3 for a step by step procedure on calculating the amperage of continuous and noncontinuous loads and sizing the OCPD based on continuous and noncontinuous operation per NEC 220-3(a).

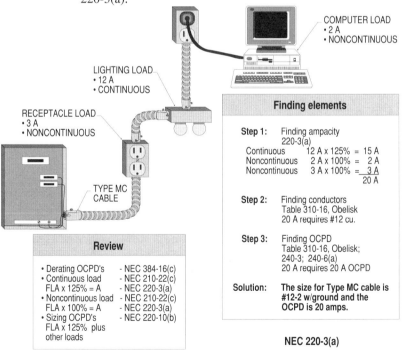

Figure 3-3. The size conductors and OCPD's shall be sized at 125% of the continuous loads plus 100% of the noncontinuous loads.

UNKNOWN LOADS

Note: It is important to remember that either the load is limited to the ampacity of the conductors in AC or MC cables or the OCPD's must protect such conductors at their allowable ampacities.

Unknown loads are loads like receptacle outlets where any size load can be easily plugged into the circuit. Unknown load circuits are required to be protected by the OCPD according to the conductor ampacity.

It has always been good design to size conductors with enough ampacity to supply the load or loads that may be connected. If the OCPD does not protect the conductors in the cable at their allowable ampacities, the load added must not exceed this value. Therefore, it is necessary to know if additional loads are likely to be connected to the circuit. **(See Note)**

See Figure 3-4 for a detailed description of adding loads or connecting loads to what the industry calls an unknown load per NEC 210-19(a) and NEC 240-3(b).

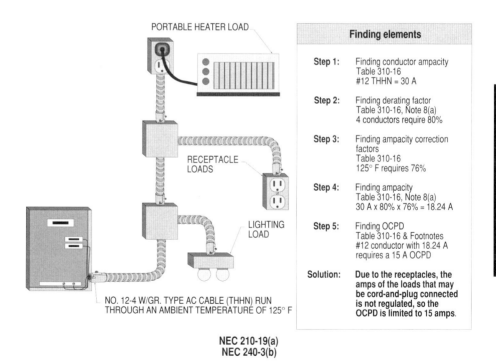

PORTABLE HEATER LOAD

RECEPTACLE LOADS

LIGHTING LOAD

NO. 12-4 W/GR. TYPE AC CABLE (THHN) RUN
THROUGH AN AMBIENT TEMPERATURE OF 125° F

	Finding elements
Step 1:	Finding conductor ampacity Table 310-16 #12 THHN = 30 A
Step 2:	Finding derating factor Table 310-16, Note 8(a) 4 conductors require 80%
Step 3:	Finding ampacity correction factors Table 310-16 125° F requires 76%
Step 4:	Finding ampacity Table 310-16, Note 8(a) 30 A x 80% x 76% = 18.24 A
Step 5:	Finding OCPD Table 310-16 & Footnotes #12 conductor with 18.24 A requires a 15 A OCPD
Solution:	**Due to the receptacles, the amps of the loads that may be cord-and-plug connected is not regulated, so the OCPD is limited to 15 amps.**

NEC 210-19(a)
NEC 240-3(b)

Overcurrent Protection

Figure 3-4. For unknown loads, the size of the OCPD must protect the conductors in the AC cable at their allowable ampacities.

KNOWN LOADS

A known load is a load that is fixed. In other words, there will be no other loads added to the circuit, such as lighting or receptacle outlets. When the ampacity of the conductors is determined and there are not any additional loads to be connected to the circuit, the OCPD does not have to match the ampacity of the conductor.

For example: An MC cable supplies a known load and has an ampacity of 17.9 amps after derating factors have been applied, the next size OCPD may be selected at 20 amps. **(See Note for unknown loads)**

See Figure 3-5 for a step by step procedure in selecting the size OCPD based upon supplying known loads per NEC 210-19(a) and 240-3(b).

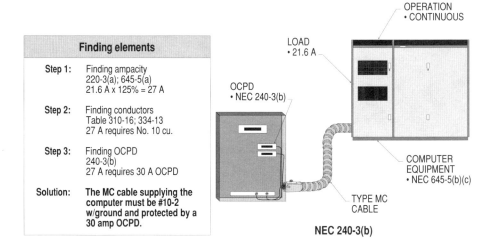

Figure 3-5. For known loads, the size of the OCPD does not have to match the ampacity of the conductors under certain conditions of use.

TERMINAL RATINGS

Overcurrent protection devices (OCPD's) and circuit terminals rated at 100 amps or less, supplied by an AC or MC cable with No. 14 through No. 1 conductors, are limited to 60°C ampacities. For overcurrent protection devices marked 60°C/75°C, the 75°C ampacities in Table 310-16 may be utilized. For circuits rated above 100 amps, with conductors larger than No. 1, 75°C ampacities of Table 310-16 are used. Ex. 1 of 110-14(c) allows higher ampacities to be used at 75°C, Ex. 2 permits conductors with higher ratings to be used if the equipment is rated and listed for higher rated conductors. The conductors rated with 90°C ampacities are only applied where the conductors require derating per Note 8(a) and correction factors per Table 310-16.

See Figure 3-6 for a step by step procedure for sizing and selecting OCPD's and conductor ampacities based upon terminal ratings per NEC 110-14(c)(1);(2).

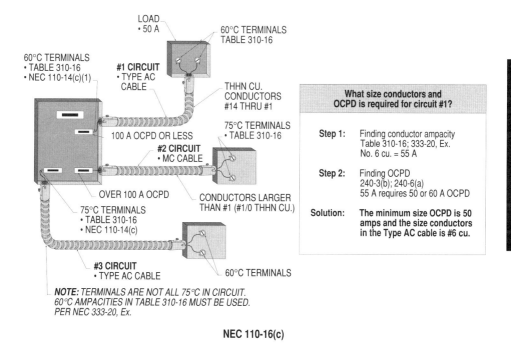

Overcurrent Protection

NOTE: TERMINALS ARE NOT ALL 75°C IN CIRCUIT. 60°C AMPACITIES IN TABLE 310-16 MUST BE USED. PER NEC 333-20, Ex.

NEC 110-16(c)

Figure 3-6. The ampacity of conductors are determined by the markings on the OCPD's and electrical equipment supplied.

NONCONTINUOUS LOADS

Type AC or MC cables supplying noncontinuous loads are computed at 100% of the total VA rating of the circuit. These loads include lighting fixtures and cord-and-plug connected equipment which are used at various intervals of time. Where such loads are properly used, they will never overload the circuit during their time of use.

For example: A feeder circuit can be used to supply a subpanel that is utilized to serve multiple branch circuit loads. Such branch circuits consist of lights, receptacles, and other utilization loads.

See Figure 3-7 for a step by step procedure for calculating the size feeder circuit used to supply noncontinuous loads per NEC 220-3(a) and 220-10(a);(b).

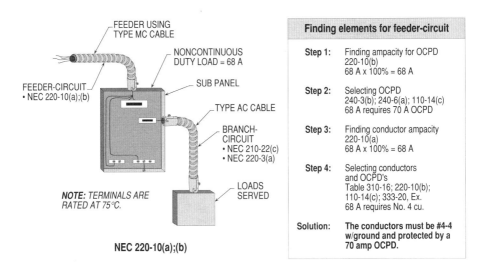

FEEDER USING
TYPE MC CABLE

NONCONTINUOUS
DUTY LOAD = 68 A

FEEDER-CIRCUIT
• NEC 220-10(a);(b)

SUB PANEL

TYPE AC CABLE

BRANCH-
CIRCUIT
• NEC 210-22(c)
• NEC 220-3(a)

LOADS
SERVED

NOTE: TERMINALS ARE
RATED AT 75°C.

NEC 220-10(a);(b)

Finding elements for feeder-circuit	
Step 1:	Finding ampacity for OCPD 220-10(b) 68 A x 100% = 68 A
Step 2:	Selecting OCPD 240-3(b); 240-6(a); 110-14(c) 68 A requires 70 A OCPD
Step 3:	Finding conductor ampacity 220-10(a) 68 A x 100% = 68 A
Step 4:	Selecting conductors and OCPD's Table 310-16; 220-10(b); 110-14(c); 333-20, Ex. 68 A requires No. 4 cu.
Solution:	**The conductors must be #4-4 w/ground and protected by a 70 amp OCPD.**

Figure 3-7. Sizing MC/AC cable to supply noncontinuous operated loads.

CONTINUOUS LOADS

Type AC or MC cables supplying continuous loads shall have their OCPD's current rating increased by 125% per NEC 220-10(b). Continuous duty loads are loads which operate for three hours or more per Article 100 of the NEC.

AC or MC cable may be used to supply such loads as lighting systems, office equipment, appliances, receptacle outlets, etc. Conductors supplying continuous duty loads sometimes have to be increased up to 125% of the load to be considered protected by the OCPD's.

See Figure 3-8 for a step by step procedure for calculating a feeder circuit used to supply continuous loads per NEC 220-3(a) and 220-10(a); (b).

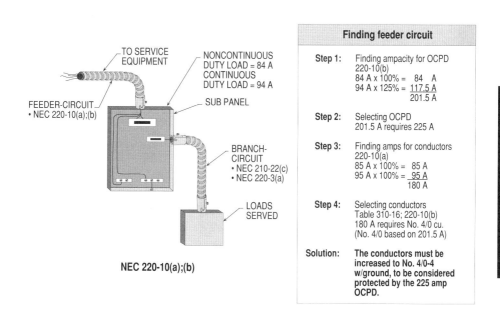

NEC 220-10(a);(b)

Finding feeder circuit

Step 1: Finding ampacity for OCPD
220-10(b)
84 A x 100% =   84   A
94 A x 125% = 117.5 A
                      201.5 A

Step 2: Selecting OCPD
201.5 A requires 225 A

Step 3: Finding amps for conductors
220-10(a)
85 A x 100% = 85 A
95 A x 100% = 95 A
                    180 A

Step 4: Selecting conductors
Table 310-16; 220-10(b)
180 A requires No. 4/0 cu.
(No. 4/0 based on 201.5 A)

Solution: **The conductors must be increased to No. 4/0-4 w/ground, to be considered protected by the 225 amp OCPD.**

Overcurrent Protection

Figure 3-8. Sizing MC cable to supply continuous loads.

Chapter Three	
Answers	**Questions**
T F	**1.** It is permissible to use the next higher standard rated device if the calculated ampacity values do not correspond to a standard OCPD rating.
T F	**2.** The next higher standard rating OCPD must be selected for circuits rated over 800 amps.
T F	**3.** Air-conditioning branch-circuit and feeders may be protected per Article 430.
T F	**4.** Motor branch-circuits and feeders may be protected per Article 430.
T F	**5.** Conductors and OCPD's shall be sized at 125% of the continuous load plus 100% of the noncontinuous loads.
T F	**6.** The size of the OCPD is not required to protect the conductors in the AC cable at their allowable ampacities for unknown loads.
T F	**7.** The size of the OCPD is required to match the ampacity for the conductors under certain conditions of use for unknown loads.
T F	**8.** A known load is a load that is fixed.
T F	**9.** Conductors rated with 90°C ampacities are only applied where the conductors require derating per Note 8(a) and correction factors per Table 310-16.
T F	**10.** Noncontinuous loads include lighting fixtures and cord-and-plug connected equipment which are used at various intervals of time.
_____	**11.** When selecting an OCPD for circuits rated over 800 amps, the next _____ standard rating shall be used.
_____	**12.** Air-conditioning branch-circuits and feeders may be protected per Article _____.

Chapter Three	
Sections	**Questions**
_____	**13.** _____ loads are loads like receptacle outlets where any size load can be easily plugged into the circuit.
_____	**14.** The ampacity of conductors are determined by the _____ on the OCPD's and electrical equipment supplied.
_____	**15.** Type AC cable supplying noncontinuous loads is computed at _____% of the total VA rating of the circuit.
_____	**16.** When sizing the OCPD for a continuous load of 14.5 amps, the load must be increased up to: **a.** 17.5 A **b.** 18.1 A **c.** 18.6 A **d.** 19.1 A
_____	**17.** To size the OCPD for a continuous load of 15.6 amps and a noncontinuous load of 4 amps, the load must be increased up to: **a.** 23.5 A **b.** 24.3 A **c.** 24.8 A **d.** 25.2 A
_____	**18.** Type AC or MC cable with No. 14 AWG through No. 1 AWG conductors with OCPD's and circuit terminals rated at 100 amps or less are limited to ampacities of: **a.** 60°C **b.** 75°C **c.** 90°C **d.** all of the above

19. Type AC or MC cable with conductors larger than No. 1 with OCPD's rated above 100 amps are required to have ampacities of:

a. 60°C
b. 75°C
c. 90°C
d. all of the above

20. Type AC cable supplying continuous loads shall have their OCPD's VA rating increased by:

a. 80%
b. 100%
c. 125%
d. 150%

INSTALLATION

There are basically two phases of rough-in work performed by electrical workers in wiring electrical systems for commercial, industrial, and residential occupancies. These phases are known in the electrical industry as mounting the boxes, enclosures, etc. and installing the wiring methods.

From the plans, electricians must locate and mount boxes for lights, switches, receptacles, and special devices including other enclosures for electrical equipment and determine the number of circuits and wiring methods required for the rough-in.

Note: All wiring methods must be installed before the slab is poured.

ROUGH-IN

The rough-in phase commences before the slab has been poured, and the interior, and exterior walls, including the ceiling have been framed.

The electrician must locate the area that the service equipment is to be installed. From this location the number of home run circuits must be determined. The service equipment is the originating point of the home run circuit wiring.

The next step is to measure and locate the boxes for the installation of receptacles, lights, and switches.

See Figure 4-1 for a detailed description of home runs using MC or AC cable to enter and leave the service equipment per NEC 300-15 and NEC 410-16.

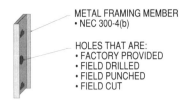

METAL FRAMING MEMBER
• NEC 300-4(b)

HOLES THAT ARE:
• FACTORY PROVIDED
• FIELD DRILLED
• FIELD PUNCHED
• FIELD CUT

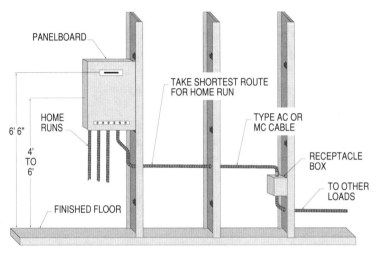

PANELBOARD

TAKE SHORTEST ROUTE
FOR HOME RUN

TYPE AC OR
MC CABLE

HOME
RUNS

6' 6"

4'
TO
6'

RECEPTACLE
BOX

TO OTHER
LOADS

FINISHED FLOOR

NEC 300-15
NEC 410-16

Review
NEC 300-4(b); (d)
NEC 333-11
NEC 333-10
NEC 333-12
NEC 334-10(f); (g)

Figure 4-1. Rough-in of AC or MC cable leaving the service equipment and entering the receptacle outlet and then routed to other loads.

MOUNTING BOXES

Note: Metal framing members are used for commercial wiring. Wood framing members are used in Figure 4-2 to illustrate the NEC rules per 300-4.

Electricians fasten the boxes to the framing members by using nails, screws, or by hammering down nails on each side of the boxes. Brackets on the sides of boxes may be hammered down to support the box.

LIGHTING OUTLETS

Metallic (metal) or nonmetallic (plastic) boxes used to support lighting fixtures are fastened to a roof or ceiling support member. A 2 in. x 4 in. framing member or bar hanger is nailed between the beams or studs and used to support the ceiling box. These hangers allow electricians to move the boxes to the exact spots called out on the plans. **Note that metal studs are used for commercial wiring.**

See Figure 4-2 for the types of mountings utilized for hanging and supporting lighting fixtures, which will be wired with armored cable per 300-15(a); (b) and 410-16(a).

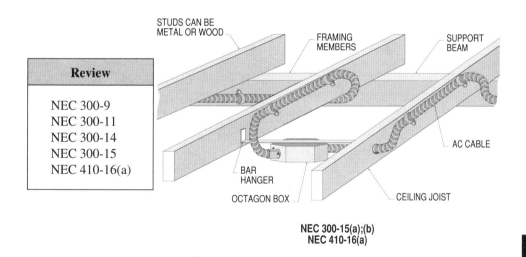

Review
NEC 300-9
NEC 300-11
NEC 300-14
NEC 300-15
NEC 410-16(a)

STUDS CAN BE METAL OR WOOD

FRAMING MEMBERS

SUPPORT BEAM

AC CABLE

BAR HANGER

OCTAGON BOX

CEILING JOIST

NEC 300-15(a);(b)
NEC 410-16(a)

Figure 4-2. Bar hangers allow electricians to move boxes and position them in the proper location where the attic space above is not accessible.

Installation

RECESSED FIXTURES

Recessed cans and lighting elements are installed in specific areas to provide special lighting effects.

This type of lighting fixture provides higher intensity light and may be aimed to specific spots. The recessed lighting can is supported by two hangers, and the electrician uses the hanger to position the fixture in the exact area that the blueprint calls for it to be.

See Figure 4-3 for a detailed description of mounting recessed cans and wiring them in with armored cable per NEC 300-15 and NEC 410-69.

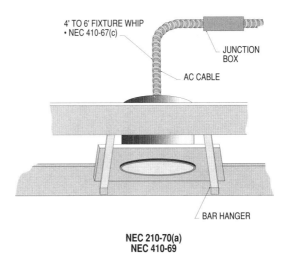

4' TO 6' FIXTURE WHIP
• NEC 410-67(c)

JUNCTION BOX

AC CABLE

BAR HANGER

NEC 210-70(a)
NEC 410-69

Review
NEC 410-65(a)
NEC 410-65(b)
NEC 410-65(c)
NEC 410-66(a)
NEC 410-66(b)
NEC 410-67(c)
NEC 410-69

Figure 4-3. Recessed lighting fixtures installed and supported with bar hangers.

Installation

SWITCH OUTLETS

Boxes supporting switches that are used to control lighting fixtures are required to be located in such a manner so that they are accessible to the user. Switches should be located at a height and location the user wishes them to be placed. However, the NEC requires switches to be located where the opening of doors will not make them inaccessible. Switches are recommended to be mounted at the following heights:

1. NEC requires not more than 6 1/2 ft. to the switch handle in the on position.
2. By experience 42 in. to 56 in. to the center of the box.

See Figure 4-4 for a detailed description of the mounting height and installation of switch outlets wired with AC & MC per NEC 380-8(a) and NEC 380-10(b).

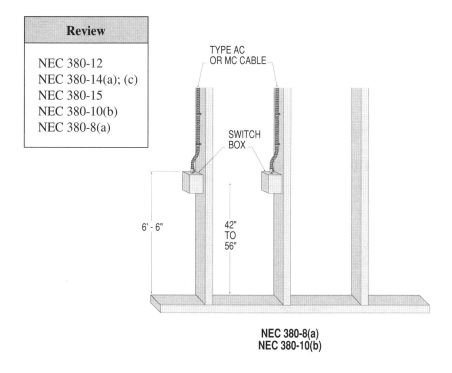

Review
NEC 380-12
NEC 380-14(a); (c)
NEC 380-15
NEC 380-10(b)
NEC 380-8(a)

NEC 380-8(a)
NEC 380-10(b)

Figure 4-4. Experience has proven that by mounting switch boxes about 42 in. to 56 in. to the center of the box will prevent the on and off handle (top) of the switch from exceeding 6 ft. 6 in. in height.

Installation

45

RECEPTACLE OUTLETS

Note: Metal framing members are used for commercial wiring. Wood framing members are used in Figure 4-5 to illustrate rules per NEC 300-4

Boxes installed on framing members are required by the NEC to be installed in designated locations to limit the use of flexible cord sets. Boxes installed to support receptacles are usually mounted at a height no greater than 5 1/2 ft. However, they are normally mounted at 12 in. centers for convenience sake. Many electricians use their hammer length or marking stick to obtain a uniform box height for receptacle boxes.

See Figure 4-5 for a detailed description of the mounting height of receptacle outlets wired with armored cable per NEC 333-9 and 334-12.

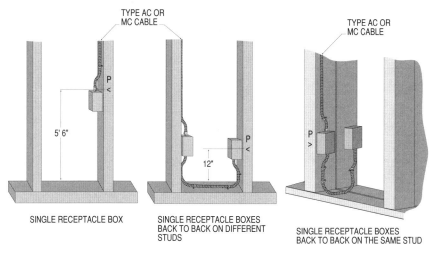

TYPE AC OR MC CABLE

TYPE AC OR MC CABLE

5' 6"

P <

P <

12"

P >

SINGLE RECEPTACLE BOX

SINGLE RECEPTACLE BOXES BACK TO BACK ON DIFFERENT STUDS

SINGLE RECEPTACLE BOXES BACK TO BACK ON THE SAME STUD

NEC 333-9
NEC 334-12

Review
NEC 333-7
NEC 333-9
NEC 333-21
NEC 334-10(a)
NEC 334-12
NEC 334-23

Figure 4-5. Studs are located and then marked for the height to mount receptacle boxes.

Installation

SPECIAL OUTLETS

Modern electrical systems must have boxes located to accommodate speakers for sound, video cameras, telephones, and computer jacks. These special outlets require a lot of additional boxes and cables to be set and routed throughout the structure of the premises. The NEC usually requires a minimum clearance of these cables from power and lighting circuit cables. **(See Note)**

See Figure 4-6 for mounting boxes for special outlets and the Type armored cable selected to wire such outlets per NEC 300-15(a); (b).

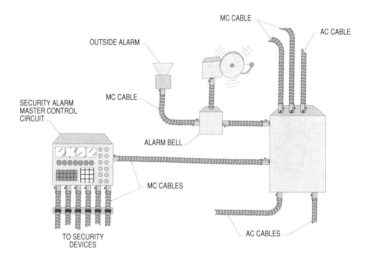

Review
NEC ART 760
NEC 820-11
NEC 820-3
NEC 820-33
NEC 820-40
NEC 820-50
NEC 820-52
NEC 820-53(d)

Figure 4-6. Special outlets are installed for the mounting of telephone and TV jacks and other types of equipment such as security alarms, etc.

Installation

DRILLING PASSAGE HOLES

Note 1: Metal studs already have holes provided for horizontal runs of cables.

Note 2: Metal framing members are used for commercial wiring. Wood framing members are used in Figure 4-7 to illustrate NEC 300-4.

After setting boxes and determining where the wiring methods are to be run, the passage holes are drilled or cut (notched) in the framing members. Properly drilled passage holes permits the cable to take the routes chosen for each run in a easy and safe manner. Cable systems are installed down from overhead or wired through other boxes and through the studs until the circuit run is complete. The holes must be drilled in the center of the 2 in. x 4 in. framing member so there is at least a 1 1/4 in. depth to help prevent a nail or screw from reaching the cable. The passage hole should be approximately 3/4 in. to 1 in. in diameter so the cable can be easily pulled without damaging the outer protective jacket. Sometimes it becomes necessary to drill from different directions to provide safe passage holes. Holes should be drilled as closes as possible to the height of the cable runs. Framing members that are notched or where the cable hole is drilled less than 1 1/4 in. from the face of the 2 in. by 4 in. stud, a 1/16 in. thick steel plate or bushing shall be used to cover the notches containing the cable to prevent the penetration of nails, screws, etc.

See Figure 4-7 for a detailed description on routing AC and MC cables through framing members per NEC 300-4(a)(1); (a)(2).

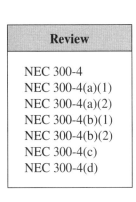

Review
NEC 300-4
NEC 300-4(a)(1)
NEC 300-4(a)(2)
NEC 300-4(b)(1)
NEC 300-4(b)(2)
NEC 300-4(c)
NEC 300-4(d)

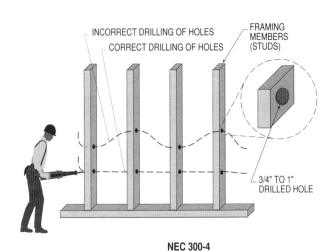

INCORRECT DRILLING OF HOLES
CORRECT DRILLING OF HOLES
FRAMING MEMBERS (STUDS)
3/4" TO 1" DRILLED HOLE

NEC 300-4

Figure 4-7. Passage holes in framing members should be drilled in a workman like manner, so that cables can be run without damaging the outer protective jacket and insulation around the conductors.

Installation

CUTTING CABLES

Note: The different methods of cutting AC/MC cables are discussed because of an industry practice. Workers are advised to use the cutting tool, which is recommended by manufacturers.

Type AC or MC cable should be cut with a special metallic-sheathed armor stripping tool. The "armor stripping tool" has many advantages. The nicking or cutting of the conductors within the metallic sheath is eliminated, plus the cuts can be made more quickly and the chance of someone accidentally cutting themselves during the cutting process is also eliminated. To make safe and reliable cable cuts, the standard practice for installers is to use the armor stripping tool recommended by most manufacturers.

See Figure 4-8 for a detailed description of armor stripping tool.

Where a "hacksaw" is utilized to cut the sheath of the armored cable, care must be taken to assure that the conductors or bonding strip are not cut or nicked. The hacksaw should be held at approximately a 60° angle to the cable armor. This procedure of cutting the cable sheath allows for a complete cutting of one spiral convolution without cutting and damaging the conductors.

A "pair of dikes" is sometimes used to cut through one spiral convolution of the armored cable and remove the cut sheath enclosing the conductors. Cable must be broken first.

Note: The above cutting methods are not recommended by AFC as an acceptable method to cut AC/MC cable.

See Figure 4-8 for a detailed description of stripping the metal sheath of armored cables.

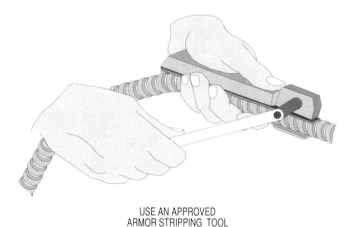

USE AN APPROVED
ARMOR STRIPPING TOOL

Figure 4-8. Method of cutting AC/MC cables, using the recommended armor stripping tool.

49

ROUTING CABLES

Note: Metal framing members are used for commercial wiring. Wood framing members are used in Figure 4-9 to illustrate NEC 300-4.

After the rough-in phase of locating and setting all boxes and enclosures, the electrician must plan and install all the cable runs. These wiring methods (cable runs) must be installed before the walls and ceilings are insulated and covered with insulation and gypsum wallboard or other wall finishing material.

With boxes in place and the panelboard enclosure installed at proper heights, wiring methods are ready to be run to the various boxes and enclosures. The electrician must determine the best and shortest routes for each circuit to be installed. Time should be taken by the electrician to plan and layout each cable run and prevent long cable runs. Long runs of spider web type routing of wiring not only requires more cable but increases voltage drop. The longer the run of wire, the more resistance that occurs and the voltage is less at the far end of the cable run.

For long runs, oversizing the circuit conductors is good practice because larger conductors will serve to limit voltage drop and assure adequate voltage at the various outlets. At centrally located junction boxes, individual three/phase, four-wire branch-circuit cables may be downsized to No. 14 on a 15 amp circuit and No. 12 on a 20 amp circuit which complies with the obelisk at the bottom of Table 310-16 of the NEC.

Sections 210-19(a), FPN 4 and 215-2(b), FPN 2 lists the methods for determining voltage drop (VD) for ungrounded conductors in branch-circuits and feeder-circuits. Section 240-23 indicates that the grounded neutral conductor may have to be increased in size due to voltage drop. Section 250-95 recognizes that equipment grounding conductors may need to be upsized due to voltage drop.

See Figure 4-9 for a detailed procedure on routing cables.

Installation

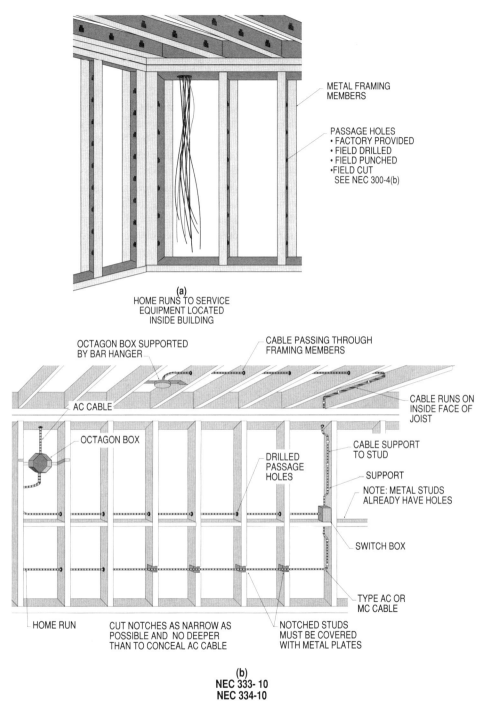

METAL FRAMING MEMBERS

PASSAGE HOLES
• FACTORY PROVIDED
• FIELD DRILLED
• FIELD PUNCHED
• FIELD CUT
  SEE NEC 300-4(b)

(a)
HOME RUNS TO SERVICE
EQUIPMENT LOCATED
INSIDE BUILDING

OCTAGON BOX SUPPORTED
BY BAR HANGER

CABLE PASSING THROUGH
FRAMING MEMBERS

AC CABLE

CABLE RUNS ON
INSIDE FACE OF
JOIST

OCTAGON BOX

DRILLED
PASSAGE
HOLES

CABLE SUPPORT
TO STUD

SUPPORT

NOTE: METAL STUDS
ALREADY HAVE HOLES

SWITCH BOX

TYPE AC OR
MC CABLE

HOME RUN

CUT NOTCHES AS NARROW AS
POSSIBLE AND NO DEEPER
THAN TO CONCEAL AC CABLE

NOTCHED STUDS
MUST BE COVERED
WITH METAL PLATES

(b)
NEC 333- 10
NEC 334-10

Figure 4-9. The proper method for routing cables through metal or wood framing members.

51

ATTICS AND BASEMENTS

Note: Metal framing members are used for commercial wiring. Wood framing members are used in Figure 4-10 to illustrate NEC 300-4.

Cables that are exposed in basements, attics, crawlspaces, etc. are required to be protected from physical damage. This protection can be provided where the cable is routed along the side of framing members such as joists, rafters, or studs. Cables run at angles to the members can be run through drilled passage holes and no further protection is required. Cables run across the bottom of framing members (joists) are required to be routed on running boards.

In accessible attics through a scuttle hole or pull down stairway, cables can run across the top of framing members or through drilled passage holes. Attic areas not accessible by permanent stairs or ladders but through a scuttle hole shall have cables protected that are run within 6 ft. of the nearest edge to the entrance area. Cables routed along the sides of the framing members or through bored passage holes do not require running boards. Cables pulled and run above 7 ft. in attic areas are considered protected from physical damage.

See Figure 4-10 for a detailed description on installing armored cables in attics and basements.

Review
NEC 333-10
NEC 334-10(f)
NEC 334-10(g)
NEC 333-11
NEC 333-12(a)

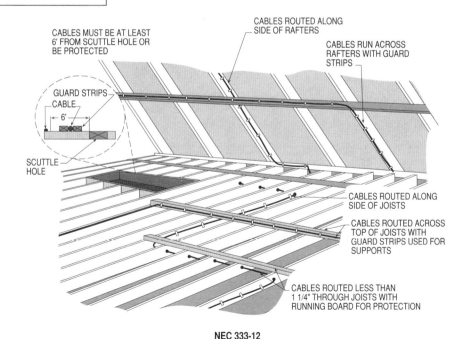

Figure 4-10. Procedures for installing cables in attic space.

Installation

EXPOSED WIRING

In general, armored cable must "hug" the surface it is wired over. Therefore, the Code prohibits bridging across open spaces. An exception is made for installations in attics, where the cable may be run without backing between joists, studs, and rafters.

There are other exceptions to this rule. In the following cases the cable may be without backing:

(a) When flexibility is required, a box may be installed near a motor or appliance, using a short piece of free cable between the box and the motor or appliance. The cable may be up to 24 in. long. This would allow flexibility in the connection, when a motor or appliance is subject to vibration.

(b) When the cable is fished, as in a partition or when the cable is run on a rack, as per Article 318, backing is not required.

(c) In basements only, where the run is not subject to physical damage, the cable may be run across the underside of the floor joists and strapped at each joist. No guard strips are required.

See Figure 4-11 for a detailed illustration of exposed wiring techniques per NEC 300-4.

PULLING CABLES THROUGH FRAMING MEMBERS

Note: Metal framing members are used for commercial wiring. Wood framing members are used in Figure 4-13 to illustrate NEC 300-4.

Cables must be pulled through the drilled framing members in a manner that won't damage the outer protective metal sheath. When pulled through wooden or metal studs, cables must be protected from the sharp or rugged edges of the punched or drilled openings. The cable is considered supported by the framing members, where it is run through the passage holes.

Care must be taken by electricians not to damage the cables when hammering the support staples to fasten the cables in place. Bends in cable runs shall be made so as not to damage the cable.

See Figure 4-12 for a detailed procedure of how to pull armored cables safely through framing members per NEC 300-4.

PULLING CABLES AROUND CORNERS

Before an armored cable can be pulled around corners of framing members, it is necessary to drill from two or even three directions to provide a route through studs. The bit must pierce both the stud facing and the one sideways, so the armored cable can pass around the corner after the holes have been drilled and completed.

See Figure 4-13 for the proper methods of drilling straight or around corner passage holes to pull cables through.

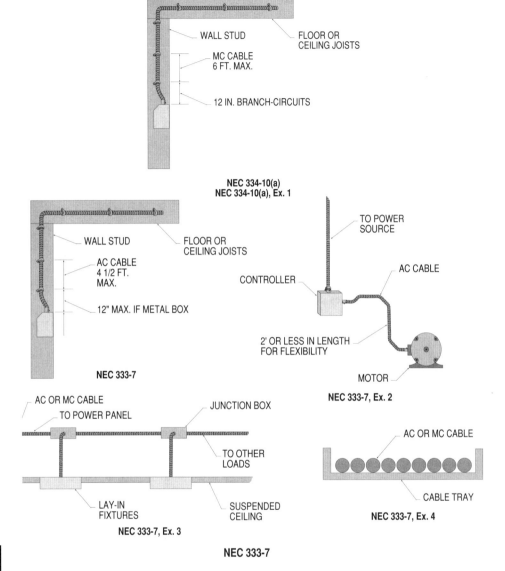

Figure 4-11. AC and MC cable can be run exposed, under certain conditions of use.

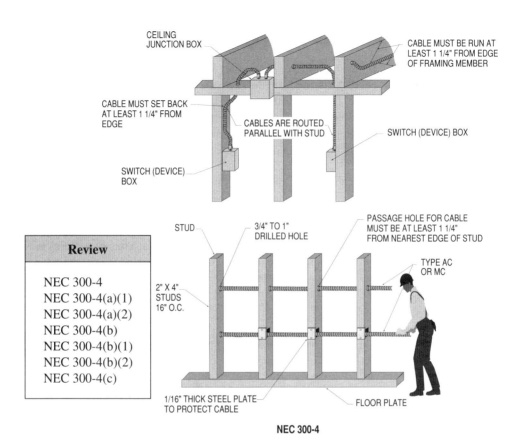

CEILING
JUNCTION BOX

CABLE MUST BE RUN AT
LEAST 1 1/4" FROM EDGE
OF FRAMING MEMBER

CABLE MUST SET BACK
AT LEAST 1 1/4" FROM
EDGE

CABLES ARE ROUTED
PARALLEL WITH STUD

SWITCH (DEVICE) BOX

SWITCH (DEVICE)
BOX

STUD

3/4" TO 1"
DRILLED HOLE

PASSAGE HOLE FOR CABLE
MUST BE AT LEAST 1 1/4"
FROM NEAREST EDGE OF STUD

TYPE AC
OR MC

2" X 4"
STUDS
16" O.C.

Review
NEC 300-4
NEC 300-4(a)(1)
NEC 300-4(a)(2)
NEC 300-4(b)
NEC 300-4(b)(1)
NEC 300-4(b)(2)
NEC 300-4(c)

1/16" THICK STEEL PLATE
TO PROTECT CABLE

FLOOR PLATE

NEC 300-4

Figure 4-12. Procedure for protecting cables in drilled passage holes or notches in wood.

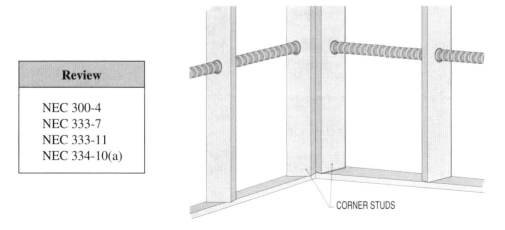

Review
NEC 300-4
NEC 333-7
NEC 333-11
NEC 334-10(a)

CORNER STUDS

Figure 4-13. AC and MC cable routed through studs and pulled around and through corner studs.

Installation

55

PREPARING AND TERMINATING CABLES

Note: Where armored cables are terminated at panelboards, switchboards, gutters, wireways, etc., the length of conductor within the enclosure is determined by the location of the splice or termination point. The cable must always be cut to allow for the additional length necessary to permit connection of the cable conductors at the desired location within the enclosure.

Before the cable or conductors can be terminated, the cable must be cut to the desired length. There are essentially two cuts that must be made. The first cut that must be taken into consideration is the length of conductor necessary to permit connection of the conductors within the enclosure. The second involves cutting of the armor to permit connection of the armor to the enclosure.

The cable sheath must be cut back to the point where the metallic sheath can be connected to the enclosure. With the sheath stripped off, the conductors are extended into the enclosure.

For example: When terminating or splicing at a junction, outlet, or switch box, the cable must be cut so that the conductors within the box are at least 6 in. in length. Therefore, the cable must be cut at least 6 in. longer for the necessary length to reach the outlet, switch, or junction box.

This additional conductor length provides enough conductor at outlets for making up joints or connections to devices. **(See Note)**

See Figure 4-14 for a detailed description of preparing and terminating AC cable.

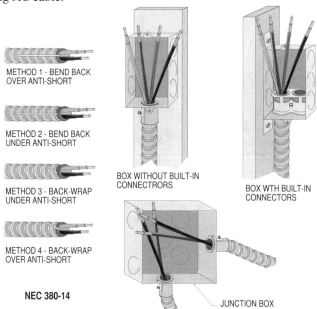

METHOD 1 - BEND BACK OVER ANTI-SHORT

METHOD 2 - BEND BACK UNDER ANTI-SHORT

METHOD 3 - BACK-WRAP UNDER ANTI-SHORT

METHOD 4 - BACK-WRAP OVER ANTI-SHORT

NEC 380-14

BOX WITHOUT BUILT-IN CONNECTRORS

BOX WTH BUILT-IN CONNECTORS

JUNCTION BOX

Figure 4-14. Procedures for preparing and terminating cables.

SUPPORTS

AC or MC cable that is run through holes or notches in framing members are generally considered to be supported even though they are not secured. The basic rule states that AC cable should be supported every 4 1/2 ft. and within 12 in. of every termination per NEC 333-7.

The general requirement for supporting MC cable differs greatly from those for AC. As covered in NEC 334-10(a), MC cable is only required to be supported every 6 ft., generally without regard for terminations at outlets, boxes, etc.

See Figure 4-15 for a detailed description of supporting AC or MC cable per NEC 333-7 and NEC 334-10(a).

There are exceptions to NEC 333-7 and NEC 334-10(a) that do not require AC and MC cables to be supported as mentioned above.

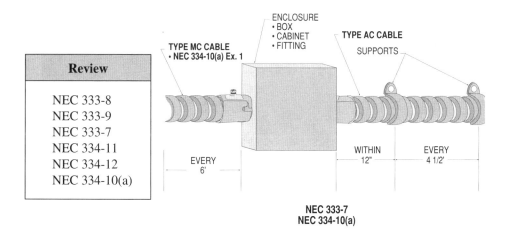

Review
NEC 333-8
NEC 333-9
NEC 333-7
NEC 334-11
NEC 334-12
NEC 334-10(a)

Figure 4-15. Procedures for supporting AC or MC cable.

57

FISHED CABLES

As previously indicated, there are situations where both AC and MC cables may be installed without supports. As recognized by Ex. 1 to NEC 333-7 and Ex. 2 to NEC 334-10(a), support requirements are waived for AC or MC, respectively, where the cables are fished. This is one of the major advantages related to the use of either AC or MC for renovation and modernization work. This flexibility allows such cables to be easily fished for significant distances as well as between joists without the use of a fish wire.

Their physical construction provides adequate mechanical strength where fished down walls and as a result, AC or MC cable can be installed more quickly and with less cutting of walls, floors, etc.

See Figure 4-16 for applications where armored cables are fished per NEC 333-7, Ex. 1 and 334-10, Ex. 2.

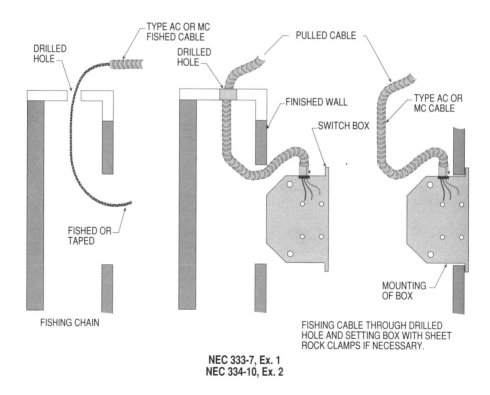

NEC 333-7, Ex. 1
NEC 334-10, Ex. 2

Figure 4-16. Methods of fishing tape and pulling cable from the attic and through the wall to mount a box and install a switch.

Installation

CABLES USED FOR FLEXIBILITY

NEC 333-7, Ex. 2 docs not require AC cable to be secured at a distance of 2 ft. or less from an outlet where flexibility is required, such as at motor terminals. Ex. 3 to NEC 333-7 recognizes unsecured lengths up to 6 ft. from outlet boxes in accessible ceilings to supply lighting fixtures. Such applications are already recognized for MC cable by the basic rule of NEC 334-10(a). However, Ex. 1 to NEC 334-10(a) mandates securing of MC cable within 12 in. of any termination where used as branch-circuits in a dwelling unit.

See Figure 4-17 for applications where armored cables are utilized for installation requiring flexibility per the NEC.

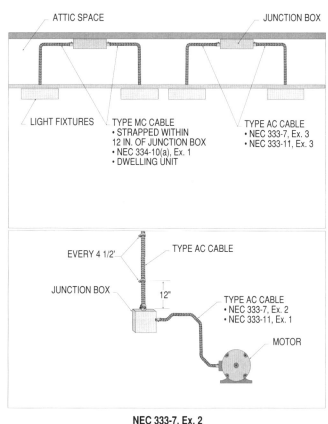

Review
NEC 333-11, Ex. 3
NEC 334-10(a), Ex. 1
NEC 333-7, Ex. 3

NEC 333-7, Ex. 2
NEC 334-10(a), Ex. 1
NEC 333-7, Ex. 3

Figure 4-17. Procedures where flexibility is utilized for the installation of equipment.

HUNG CEILINGS

NEC 300-11(a) generally prohibits the use of "support wires" as the sole means of support for cables where the support wires "do not provide secure support." However, NEC 300-11(a) does recognize the use of hung ceiling support wires as the sole means of support for cables, boxes, etc. provided certain conditions are satisfied. Such permission is limited to branch-circuit conductors that supply equipment "located within, supported by, or secured to a non-fire-rated floor or roof/ceiling assembly." NEC 300-11(a) recognizes support wires as the sole means of support for branch-circuits supplying surface and recessed lighting in the ceiling and their associated switches, wiring for floor-to-ceiling power poles and columns containing receptacles, as well as communication and data outlets.

See Figure 4-18 for applications using support wires of suspended ceilings (hung ceilings) to support armored cables per NEC 300-11(a).

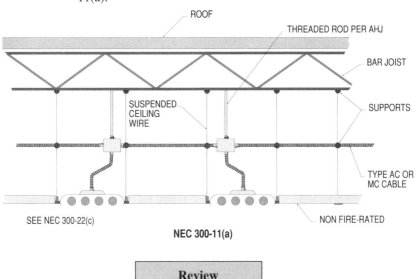

NEC 300-11(a)

Review
NEC 300-11(a)
NEC 370-23(b)
NEC 410-16(c)
NEC 333-7, Ex. 3

Figure 4-18. AC and MC cables may be supported to suspended ceiling wires under certain conditions of use.

USING RACEWAYS FOR SUPPORTS

The provisions of NEC 300-11(b) generally prohibits the use of raceways as the means of support for cables or nonelectrical equipment. Additionally, AC and MC cable may not be used to support other electrical conductors, cables, and equipment or nonelectrical cables and equipment. Again, there are exceptions. As given in Ex. 1, large conduits equipped with hangar bars or fittings intended to support smaller cables are permitted to be used. Ex. 2 recognizes the support of Class 2 control cables by the conduit carrying the power conductors supplying the associated, controlled equipment. Armored cables are required to be secured by "approved" staples, straps, hangers, etc. Although, the term "approved" is defined as that which is acceptable to the authority having jurisdiction (AHJ).

Note: No maximum distance is given for securing AC cable to vertical runs for cable tray. Usually the required support per NEC 333-7 is sufficient.

See Figure 4-19 for applications of using raceways to support cables.

USING CABLE TRAYS FOR SUPPORTS

Both AC and MC cables are permitted to be installed in cable tray. For cables installed in cable tray both AC and MC are required to be secured only in vertical runs as dictated by NEC 318-8(b). **(See Note)**

Review
NEC 300-11(b)
NEC 300-11(b), Ex. 1
NEC 300-11(b), Ex. 2
NEC 300-11(b), Ex. 3

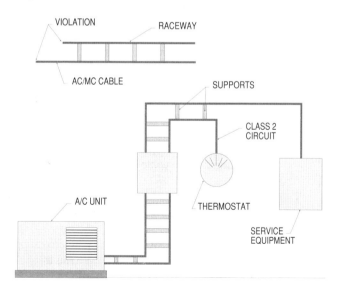

NEC 300-11(b)
NEC 300-11(b), Ex. 2

Figure 4-19. AC and MC cables are permitted to be supported by raceway systems when they are Class 2 circuits per 300-11(b).

Installation

Chapter Four	

Answers	Questions
T F	**1.** A bar hanger allows electricians to move boxes and position them in the proper location.
T F	**2.** Type AC cable is allowed to be installed in 4 ft. to 7 ft. fixture whips for recessed cans.
T F	**3.** The switch handle in the on position is required to be installed at a height not exceeding 5 1/2 ft.
T F	**4.** Boxes installed to support receptacles are usually mounted at a height no greater than 6 1/2 ft.
T F	**5.** Boxes located to accommodate speakers for sound, video cameras, telephones, and computer jacks are mounted at heights that accommodate their conditions of use.
T F	**6.** Passage holes should be approximately 3/4 in. to 1 in. in diameter so the cable can be easily pulled without damaging the outer protective jacket of Type AC or MC cable.
T F	**7.** Passage holes must be drilled in the center of the 2 in. x 4 in. framing member so there is at least 1 1/2 in. depth to help prevent a nail or screw from reaching the cable.
T F	**8.** A cutting tool is recommended by most manufacturers to make safe and reliable cuts.
T F	**9.** Long runs of spider web type routing of wiring not only requires more cable but increases voltage drop.
T F	**10.** Cables that are exposed in basements, attics, crawlspaces, etc. are not required to be protected from physical damage.
T F	**11.** Type AC or MC cable is considered supported by the framing members where they are run through the passage holes.
T F	**12.** When terminating or splicing at a junction, outlet, or switch box, the cable must be cut so that the conductors within the box are at least 6 in. in length.

	Chapter Four
Answers	**Questions**
T F	**13.** Type AC cable is required to be supported every 5 1/2 ft. and within 12 in. of every termination.
T F	**14.** Type MC cable shall be secured within 24 in. of any termination when used as branch-circuits in a dwelling unit.
T F	**15.** Both Type AC and MC cables are permitted to be installed in cable trays.
_____	**16.** Switches are required to be mounted to the switch handle in the on position at _____ ft.
_____	**17.** Boxes installed to support receptacles are required to be mounted at a height no greater than _____ ft.
_____	**18.** Receptacles are usually mounted at _____ in. centers for convenience sake.
_____	**19.** Framing members that are notched or where the cable hole is drilled less than 1 1/4 in. from the face of the 2 in. x 4 in. stud, a _____ in. thick steel plate is required.
_____	**20.** When cutting the cable armor of Type AC or MC cable with a hacksaw, the cable should be cut at a _____ angle.
_____	**21.** Attic areas that are accessible through a scuttle hole shall have cables protected from the nearest edge to the entrance area at least: **a.** 3 ft. **b.** 4 ft. **c.** 5 ft. **d.** 6 ft.
_____	**22.** Cables pulled and run in attic areas are considered protected from physical damage when installed above: **a.** 5 ft. **b.** 6 ft. **c.** 7 ft. **d.** 10 ft.

Sections	Questions
_____	**23.** Type MC cable is only required to be supported every: **a.** 3 ft. **b.** 4 1/2 ft. **c.** 5 ft. **d.** 6 ft.
_____	**24.** Where flexibility is required, Type AC cable is not required to be supported in lengths up to: **a.** 1 ft. **b.** 2 ft. **c.** 3 ft. **d.** 6 ft.
_____	**25.** Type AC cable used in accessible ceilings to supply lighting fixtures from outlet boxes can be installed in unsecured lengths up to: **a.** 6 ft. **b.** 7 ft. **c.** 8 ft. **d.** 10 ft.

SPECIAL CIRCUITS AND EQUIPMENT

Special circuits and equipment are characterized by their power limitations and condition of use that differentiate them from electric lighting and power circuits. The wiring methods, grounding requirements, protection rules and types of conductors employ a wide variety of installation techniques. Special types of equipment and materials designed for specific applications are usually utilized.

The various types of armored cables can be utilized to supply clean and reliable power to such circuits and equipment.

ELECTRONIC COMPUTER AND PROCESSING CIRCUITS

The electrical environment for computers includes their power sources, grounding and electrical interfaces with communication lines, air conditioning, and life safety systems. It also includes lighting and other non-computer electrically operated equipment in and about the computer area.

The electrical environment immediately located around computers must also be considered, since electrical disturbances propagate through conductors, pipes, metal ducts and conductive structural members or by radiation such as radio waves. External sources of electrical disturbances range from lighting to nearby radio transmitters and electrical loads which generate electrical noise when operated or switched on and off. Internal sources of electrical disturbances within the computers may generate electrical noises and disturbances. Therefore, clean and reliable power must be provided for such systems.

Custom-designed armored cables specially produced for home runs between the power source and computers are available from AFC. These are available in any combination of insulated conductors with printing, or stripping to meet specific design and installation needs.

COMPUTER GROUNDING

Grounding accomplishes many functions, all of which must be considered in the design and installation of any computer circuits. Personnel concerned with only one or two of these functions may violate others in ignorance.

For example: Grounding is required for the safety of personnel and to protect the computer from unwanted electrical signals and noise. AC and MC cables may be provided with more than one grounding conductor to ground highly sensitive computers in such a manner so they will operate reliably. There are two methods in which computers are wired. They are hard wired or cord-and-plug connected.

CORD-AND-PLUG CONNECTED

Sensitive electronic equipment such as computers are sometimes cord-and-plug connected to obtain their power supply. Computers sharing the same branch-circuits may interfere with each other.

For example: A small motor type appliance on the same circuit as a computer may cause power line transients which can upset the digital processor. One solution is to run MC cables with individual branch-circuits to reduce such noise couplings. If noise problems still exist, it is better to run MC cables with separate grounding conductors and install isolated/insulated grounding (IG) receptacles.

The computer plugged into this form of receptacle is grounded at a specific upstream point rather than locally at the outlet box through a bonding jumper. This demands a separate equipment grounding conductor in the AC and MC cable to connect the receptacle grounding pole to this remote upstream point. This upstream point can be any subfeed panel or the grounding bus at the service equipment. This IG conductor may be run through one or more panelboards before it is connected to the grounding electrode system.

See Figure 5-1 for a detailed description of cord-and-plug connecting sensitive electronic equipment such as computers per NEC 250-74, Ex. 4 and NEC 384-20, Ex.

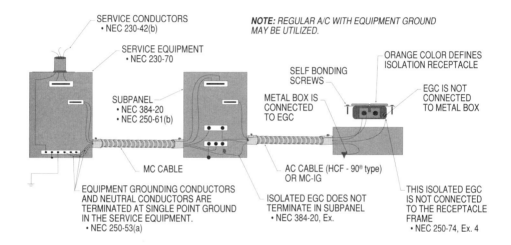

SERVICE CONDUCTORS
• NEC 230-42(b)

SERVICE EQUIPMENT
• NEC 230-70

NOTE: REGULAR A/C WITH EQUIPMENT GROUND MAY BE UTILIZED.

ORANGE COLOR DEFINES
ISOLATION RECEPTACLE

SELF BONDING
SCREWS

SUBPANEL
• NEC 384-20
• NEC 250-61(b)

METAL BOX IS
CONNECTED
TO EGC

EGC IS NOT
CONNECTED
TO METAL BOX

MC CABLE

AC CABLE (HCF - 90® type)
OR MC-IG

EQUIPMENT GROUNDING CONDUCTORS
AND NEUTRAL CONDUCTORS ARE
TERMINATED AT SINGLE POINT GROUND
IN THE SERVICE EQUIPMENT.
• NEC 250-53(a)

ISOLATED EGC DOES NOT
TERMINATE IN SUBPANEL
• NEC 384-20, Ex.

THIS ISOLATED EGC
IS NOT CONNECTED
TO THE RECEPTACLE
FRAME
• NEC 250-74, Ex. 4

NEC 250-74, Ex. 4
NEC 384-20, Ex.

Figure 5-1. Using MC cable to supply a subpanel and HCF - 90® type cable to serve an isolation receptacle. ***Note:*** *Two EGC's are run to isolation receptacle and box.*

HARD WIRED

Sensitive electronic equipment may be hard wired using Type MC/IG cable to route power between the power source and equipment to be served, and permitting a nonmetallic spacer or fitting between the MC cable and enclosure of the equipment. The spacer or fitting is installed to reduce the electrical noise (electromagnetic interference) that may interfere with the operation of sensitive electronic equipment. However, the MC/IG cable must be grounded and bonded at one end and an equipment grounding conductor must be used to ground the sensitive electronic equipment such as a computer.

See Figure 5-2 for a detailed description of hard wiring a sensitive piece of electronic equipment using MC cable per NEC 250-75, Ex. and NEC 384-20, Ex.

COMPUTER NEUTRALS

In 208/120 V wye systems, where load currents are rich in third and higher harmonic content, the neutral currents cannot cancel in zero sequence fashion. The result can be excessive voltage drop which produce wave form distortion and overheating where the neutral is not of sufficient size. Larger neutral conductors with greater cross section area may be used and usually reduces and eliminates this problem. Another method to help cure the problem is to provide at least one full-size individual neutral for each phase conductor which is routed to each computer.

Super Neutral Cable is a metal clad, MC cable, manufactured with an oversized neutral conductor or one neutral per phase for three/phase, four-wire power supply systems to computers (with DC drive fan motors, tape, and disk drives), office machines, programmable controls, and similar electronic equipment where non-linear switching loads produce additive, third order harmonic current, which may create overloaded neutral conductors.

The oversized neutral conductor(s) are sized 150% to 200% of the phase conductor ampacity to minimize the effects of harmonics generated by the non-linear loads. The neutral per phase (striped with color to match the phase conductor) accomplishes the same objective.

See Figure 5-3 for applications using MC cable with custom-designed neutrals to supply computer systems.

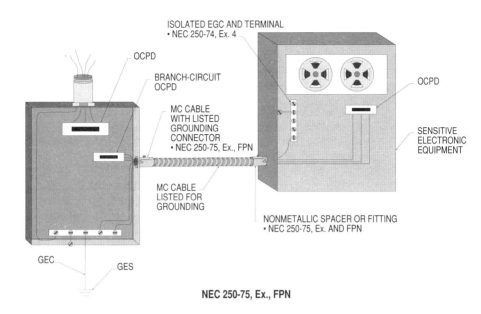

ISOLATED EGC AND TERMINAL
• NEC 250-74, Ex. 4

OCPD

BRANCH-CIRCUIT
OCPD

MC CABLE
WITH LISTED
GROUNDING
CONNECTOR
• NEC 250-75, Ex., FPN

MC CABLE
LISTED FOR
GROUNDING

OCPD

SENSITIVE
ELECTRONIC
EQUIPMENT

NONMETALLIC SPACER OR FITTING
• NEC 250-75, Ex. AND FPN

GEC

GES

NEC 250-75, Ex., FPN

Figure 5-2. MC cable, listed for grounding, may be isolated at sensitive electronic equipment by using a nonmetallic space or fitting.

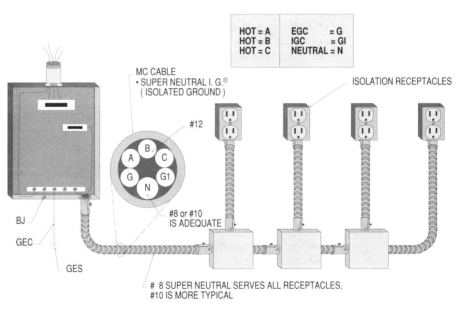

HOT = A	EGC	= G
HOT = B	IGC	= GI
HOT = C	NEUTRAL = N

MC CABLE
• SUPER NEUTRAL I. G.®
  (ISOLATED GROUND)

ISOLATION RECEPTACLES

#12

#8 or #10
IS ADEQUATE

BJ

GEC

GES

8 SUPER NEUTRAL SERVES ALL RECEPTACLES,
#10 IS MORE TYPICAL

NOTE 10, PART C TO TABLES 310-16

Figure 5-3. MC cable with custom designed super neutral can be used to serve a number of hot branch-circuit conductors supplying isolation receptacles.

FIRE ALARM CIRCUITS

Fire-related alarm circuit conductors are generally required to be run separately in cables containing only those conductors, which are sized and selected per the NEC.

Interfaces must not be permitted to create unwanted ground loops capable of conducting electrical noise into the grounding system. This can be controlled partly by providing the single entry point principle of all wiring methods.

Electrical noise, often present on the conductors to remotely located sensors, can be controlled by various methods appropriate to the type of signal to be transmitted and the characteristics of major noise sources involved. Properly designed phase, neutral, and grounding conductors help alleviate such noise problems.

Dual Rated Fire Alarm/Control Cables are temperature rated 105°C for power limited fire-protective signaling cable (Type FPLP) applications and 90°C for non-power limited circuit cable (Type MC) installations. FPLP cable is fully plenum rated, low smoke, and fire resistant.

ALARM CIRCUITS

One of the most critical components in any fire alarm system is cables run above suspended ceilings, within the walls or partitions, and under floors. This electrical system interconnects pull stations, smoke detectors, alarm devices, bells, horns, and other equipment to the main fire alarm control panel.

Red armored fire alarm/control cable is the preferred replacement for pipe and wire in fire alarm runs. These cables are listed and labeled by UL for such use.

Fire alarm/control cable is competitive with Teflon® cables and gives contractors a choice in sizes. Contractors will find a wide selection of connectors available from most manufacturers.

Red armored type MC cables are equipped to be installed in plenums, spaces used for environmental air and places of public assembly per NEC 300-22(c) and NEC 518-4, and NEC 760-51(d).

REMOTE CONTROL CIRCUITS

In addition to its use for fire alarm wiring, a red armored cable assembly is the ideal answer for any and all runs of remote-control circuits from magnetic motor starters, contactors, relays, and signaling devices. It also assures fast, flexible connection to push button stations, float switches, limit switches, pressure switches, and all other types of pilot control devices, as permitted by NEC 430-72, NEC 725-12 and 16.

Red alarm/control cable is designed to provide the same kind of time saving, cost saving, labor and materials saving as armored cable, Red armored fire alarm/control cable offers designers and installers a high quality, cost-effective replacement for pipe and wiring systems in all public and private buildings.

Conductors are copper, solid sizes No. 18. Through No. 12 AWG conductors are available from AFC. Twisted pairs and twisted shielded pairs are now commonly stocked items. Combinations of pairs and single conductors custom-designed are available by special inquiry.

An internal copper equipment grounding conductor is provided in each cable and can be specified either bare or insulated.

See Figure 5-4 for an application using red armored fire alarm/control cable to wire in fire alarm circuits per UL 1569.

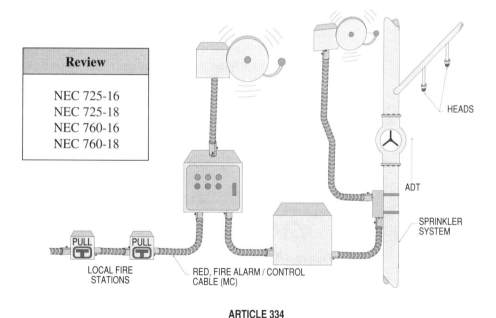

ARTICLE 334
UL 1569

Figure 5-4. Red fire alarm/control MC cable can be used to supply circuits for fire alarm circuits.

COMMUNICATION CIRCUITS

Communication circuit conductors vary in size, type of insulation, and combination of conductors, depending upon the system requirements. They may be installed in a cable assembly such as AC or MC which often contains a large number of small sized identifiable conductors. Conductors are connected directly to the equipment or to terminal blocks.

Communication and signal circuits must be installed in a separate cable from power circuits and both must be grounded in accordance with the rules listed in the NEC. Communication and power conductors must not enter the same outlet boxes unless the conductors are separated by a partition or the power circuit conductors are introduced solely for power supply to communication equipment or for connection to remote-control equipment.

SMOKE DETECTOR CIRCUITS

Ordinances and building codes require smoke detectors to be installed in all new homes, apartment complexes, and some commercial buildings.

Smoke detectors should be placed in the hallways, on each floor level, near doorways where bedrooms are located, in the basements, kitchens and even in the attic. Locations of smoke detectors are dictated by building codes and local ordinances. Smoke detectors shouldn't be considered luxuries, but necessities. AC or MC cables make excellent wiring methods to wire-in outlets servicing such detectors.

The outer metal sheath provides the protection necessary to protect the circuit conductors from physical damage and abuse and maintain reliable and dependable power to the smoke detector continuously.

See Figure 5-5 for a detailed description of locations where smoke detectors are required to be installed.

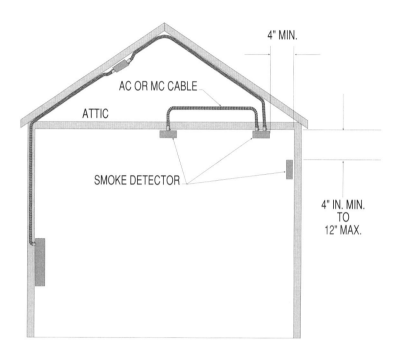

Figure 5-5. AC or MC cable may be used to supply power to smoke detectors. Check with local codes for location requirements.

SECURITY CIRCUITS

Regarding security system installations, a telephone extension cable, which consists of No. 22 to No. 24 conductors, which connects the various magnetic switches and devices in such a system is utilized. The supply conductors can be obtained in a pair in an armored cable which can easily be routed through the attic space and down the inside walls and terminated.

AC and MC cable are excellent wiring methods to route branch-circuits to supply outlets providing outside security lighting mounted on the outside walls of buildings. The armored cable provides the necessary protection of circuit conductors to ensure that power is maintained continuously to lighting fixtures so that proper lighting is always provided.

See Figure 5-6 for accepted methods of wiring in security lights with AC or MC cables.

INTRINSICALLY SAFE CIRCUITS

MC cable can be utilized to wire in low voltage circuits in locations that are classified as hazardous. Such circuits are found only in process control and similar low energy circuits. Power circuits used to drive electrical motors, provide lighting or operate large equipment, are not suitable for intrinsically safe applications due to the amount of power required. Class I, Division 2 equipment including suitable wiring methods such as MC cable must be used. Accessories used with such wiring methods and equipment must be listed and labeled as explosion proof in most all installations.

AC or MC cable may be used to wire intrinsically safe circuits supplying low energy power to instruments installed in the process industry, where flammable atmospheres may be present. Such low voltage instrumentation is thermocouples, RTD's, LTD's, and proximity switches which are of low-energy design, suitable for hazardous areas.

See Figure 5-7 for a detailed illustration of how to wire such instruments with AC or MC cable per NEC 504-20.

MC cable is approved to wire-in equipment installed in Class II, Division 2 and Class III, Division 1 and 2 locations. Section 502-4(b) allows MC cable to be installed as a wiring method in Class II, Division 2 locations, where combustible dust may be present. Section 503-3(a) and (b) recognizes MC cable as a wiring method to wire in equipment installed in Class III, Division 1 and 2 locations, where easily ignitable fibers or flyings are present.

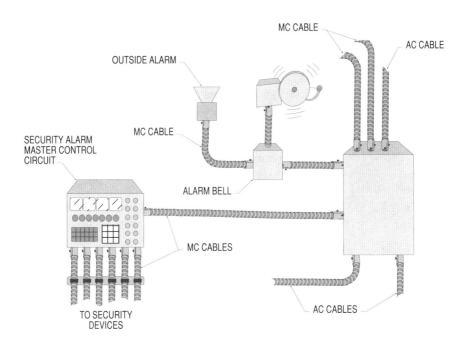

Figure 5-6. AC and MC cables can be used to wire-in elements of security systems.

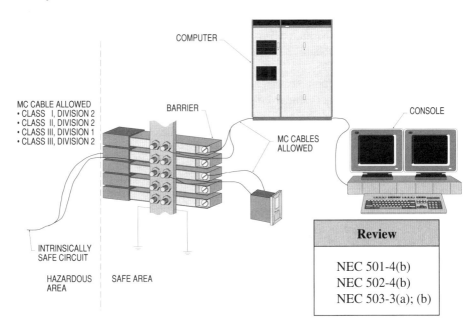

Figure 5-7. MC cables may be used in the safe area to wire-in intrinsically safe circuits.

75

Chapter Five	
Answers	**Questions**
T F	**1.** Custom-designed armored cables for home runs between the power source and computers are available from AFC.
T F	**2.** Armored cables are equipped with more than one grounding conductor to ground highly sensitive computers in such a manner so they will operate reliably.
T F	**3.** Sensitive electronic equipment such as computers are never cord-and-plug connected to obtain their power supply.
T F	**4.** Sensitive electronic equipment may be hard wired using Type MC cable.
T F	**5.** Super Neutral cable is a metal clad, Type MC cable, manufactured with an undersized neutral conductor.
T F	**6.** Red armored Type MC cables are not required to be installed in spaces used for environmental air and places of public assembly.
T F	**7.** Red armored Type MC cables are available in solid sizes No. 18 through 12 AWG.
T F	**8.** Communication and signal circuits must be installed in a separate cable from power circuits.
T F	**9.** Location of smoke detectors are dictated by electrical codes.
T F	**10.** Type MC cable shall not be used to wire intrinsically safe circuits.
_____	**11.** Computers sharing the _____ may interfere with each other.
_____	**12.** Oversized neutral conductors are sized _____ to _____ of the phase conductor ampacity.
_____	**13.** Red armored Type MC cables are available in solid sizes _____ thru _____.

Answers	Questions
_____	**14.** Type AC or MC cables can be used to wire-in elements of _____ systems.
_____	**15.** MC cables may be used in the safe areas to wire-in _____ safe circuits.
_____	**16.** Red armored Type MC cables are available by special inquiry in more than _____ conductors. **a.** 8 **b.** 6 **c.** 4 **d.** 2
_____	**17.** Ordinances and building codes require that smoke detectors be installed in all new: **a.** homes **b.** apartment complexes **c.** commercial buildings **d.** all of the above
_____	**18.** Regarding security systems, a telephone extension cable shall consist of conductors in sizes: **a.** No. 22 to No. 24 **b.** No. 18 to No. 22 **c.** No. 16 to No. 18 **d.** No. 14 to No. 16
_____	**19.** Where combustible dust may be present, Type MC cable shall be installed in: **a.** Class I, Division 2 **b.** Class II, Division 1 **c.** Class II, Division 2 **d.** Class III, Division 1

20. Where easily ignitable fibers or flyings are present, Type MC cable shall be installed in:

a. Class I, Division 1 and 2
b. Class II, Division 1 and 2
c. Class III, Division 1 and 2
d. none of the above

SPECIAL FACILITIES AND LOCATIONS

Armored cables such as AC and MC are ideal for use as wiring methods for special facilities and locations where reliability is the major concern, maximum performance is demanded, space is limited and ease of application is critical. Advanced cable systems offer particular advantages over raceway and conductor systems. A cable system offers contractors a cost-efficient alternative to raceway and conductor systems.

For example: Cable pullers, benders, wire fish tapes, pipe threaders, reamers, wire dispensers, lubricants, and elbows are eliminated. Also gone are the materials, tools, and labor intensive operations required to install such wiring methods as EMT, IMC, rigid metal conduit, and flexible metal conduit. The savings above can be 40 - 50% over EMT and conductors and up to 30 - 40% over flexible metal conduit and conductors.

OTHER SPACES USED FOR ENVIRONMENTAL AIR

Other spaces used for environmental air are spaces used for environmental air-handling purposes other than ducts and plenums. Such spaces are above suspended ceilings or finished ceilings which are covered with sheet rock material, etc. These spaces are restricted to certain types of wiring methods per NEC 300-22(c) and NEC 760-51(d).

For example: AC and MC cables are permitted to be utilized as wiring methods for electrical equipment located in such spaces.

AFC's Red Fire Alarm/Control Cable has passed and complies with UL 910 Steiner Tunnel Test and is fully plenum rated as Type FPLP/MC.

Red Fire Alarm / Control Cable is suitable for use in ducts, plenums, and other spaces used for environmental air. It is listed as fully plenum rated and having fire resistant and low-smoke characteristics.

See Figure 6-1 for a detailed application of installing Red Fire Alarm / Control Cable in spaces used for environmental air.

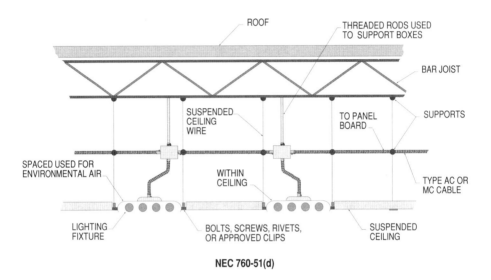

NEC 760-51(d)

Review
UL 910
UL 1569
UL 1424
NEC 300-22(c)
NEC 370-23(b)
NEC 410-16(c)
NEC 410-67(c)
NEC 760-51(d)

Figure 6-1. MC Red Fire Alarm/Control Cable may be used as a wiring method in spaces used for environmental air and plenums.

HAZARDOUS LOCATIONS

Hazardous locations are areas where combustible and flammable materials are present or may be present during operation of electrical equipment. To prevent explosions from occurring, wiring methods and equipment are limited to certain types. Selection of wiring methods and equipment are based upon Class I, II, or III locations and Divisions of 1 or 2.

MC cable is permitted as a wiring method in Class I, Division 2 locations containing volatile flammable liquids or flammable gases. Such locations are refineries, chemical plants, etc.

MC cable can be used as a wiring method in locations containing dust and areas having flyings and fibers.

CLASS I, DIVISION 2 LOCATIONS

Notice that armored cable is not recognized for use in Class I hazardous locations for general wiring, but MC is permitted for use in Class I, Division 2 hazardous locations for the wiring of intrinsically safe equipment. Such equipment consists of instruments or signals in which the electric circuit is not capable of releasing enough energy under any condition to cause ignition of the hazardous atmosphere.

See Figure 6-2 for MC cables used to wire intrinsically safe circuits and equipment in Class I, Division 2 locations per NEC 504-20.

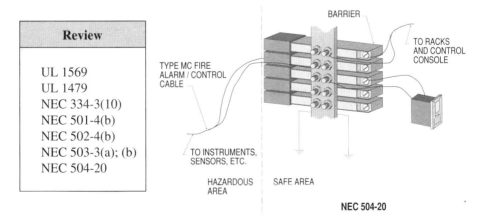

Review

UL 1569
UL 1479
NEC 334-3(10)
NEC 501-4(b)
NEC 502-4(b)
NEC 503-3(a); (b)
NEC 504-20

BARRIER

TO RACKS
AND CONTROL
CONSOLE

TYPE MC FIRE
ALARM / CONTROL
CABLE

TO INSTRUMENTS,
SENSORS, ETC.

HAZARDOUS
AREA

SAFE AREA

NEC 504-20

Figure 6-2. MC fire alarm/control cable may be used in hazardous locations to supply low-energy power to instruments, sensors, etc.

MC cable is allowed to be utilized as a wiring method in Class I, Division 2 locations to wire-in various types of equipment. MC cable may be installed in cable trays, if installed in such a manner to avoid tensile stress at termination points.

See Figure 6-3 for application of MC cable used as a wiring method in Class I, Division 2 locations per NEC 501-4(b).

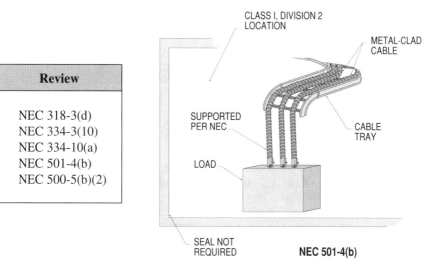

Review

NEC 318-3(d)
NEC 334-3(10)
NEC 334-10(a)
NEC 501-4(b)
NEC 500-5(b)(2)

Figure 6-3. MC cable may be used in hazardous locations and installed in cable trays.

Note: AC or MC cables may be used to wire nonincendive circuits and equipment such as low-energy signals for instruments.

CLASS II, DIVISION 2 LOCATIONS

MC cable is an excellent wiring method to wire-in equipment located in Class II, Division 2 locations. MC cables, if installed in a single layer may be installed in cable tray systems. However, ladder type, ventilated trough, or ventilated channel cable trays must be used. **(See Note)**

See Figure 6-4 for AC and MC cables installed in Class II, Division 2 locations per NEC 502-4(b).

CLASS III, DIVISION 1 AND 2 LOCATIONS

Type MC cable is permitted as a wiring method for all types of electrical equipment installed in Class III, Division 1 and 2 locations. The size of such cables must be selected with allowable ampacity ratings to prevent overheating the surface temperature of the equipment.

For example: A piece of equipment pulling 15.5 amps must be supplied with a No. 12-2 with ground, MC cable, rated at 20 amps. The No. 12 conductors in the cable, if they are copper, have enough ampacity to supply the 15.5 amp load and prevent overheating.

See Figure 6-5 for applications using MC cables in Class III, Division 1 and 2 locations per NEC 503-3(a);(b)

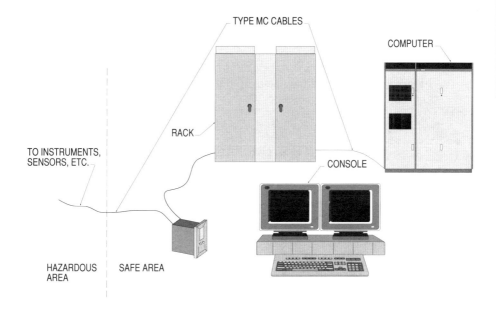

Figure 6-4. MC Cable may be used for nonincendive circuits that supply low-energy circuits to instruments, etc. located in Class II, Division 2 locations.

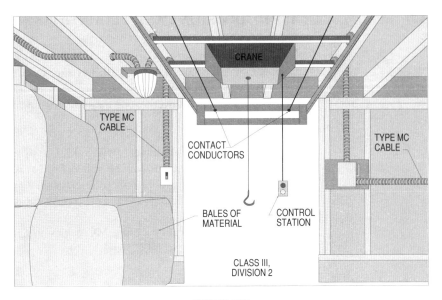

NEC 503-3(b)

Figure 6-5. MC Cable may be used to wire equipment located in Class III, Division 1 and 2 locations.

HEALTH CARE FACILITIES
517-13(a), Ex. 1

Wiring methods used in health care facilities must be dependable and reliable to provide the necessary power needed for the operation of equipment used for the caring of patients.

There are a number of AC and interlocked-armor MC cables that are intended to be used as wiring methods in health care facilities.

For example: AC cable does not normally have an equipment grounding conductor. However, AC cable is available with up to four conductors, plus it may include an equipment grounding conductor.

Such cables are commonly referred to as HCF-90 or "Hospital Grade" Type AC.

See Figure 6-6 for an illustration of HCF-90, "Hospital Grade" Type AC cable per UL 4 and UL 1479.

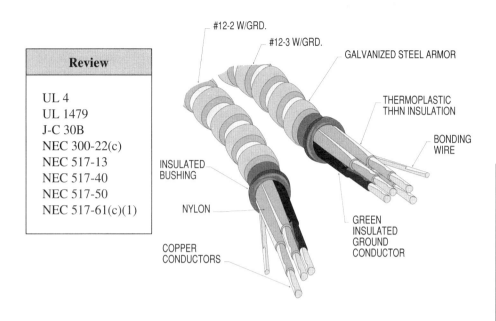

<table>
<tr><td>

Review

UL 4
UL 1479
J-C 30B
NEC 300-22(c)
NEC 517-13
NEC 517-40
NEC 517-50
NEC 517-61(c)(1)

</td></tr>
</table>

#12-2 W/GRD.

#12-3 W/GRD.

GALVANIZED STEEL ARMOR

THERMOPLASTIC
THHN INSULATION

BONDING
WIRE

INSULATED
BUSHING

NYLON

COPPER
CONDUCTORS

GREEN
INSULATED
GROUND
CONDUCTOR

Special Facilities & Loc.

Figure 6-6. AC Cable, which is specified as HCF-90 (for use in health care facilities) may be used in such locations.

REDUNDANT GROUNDING

The purpose of an extra equipment grounding conductor is to provide the redundant grounding required at patient bed locations per NEC 517-13(a), Ex. 1.

All metal electrical equipment operating at over 100 volts in areas that are or may be occupied by patients and which are subject to contact by persons must be grounded by an insulated copper conductor sized in accordance with Table 250-95 of the NEC. The copper grounding conductor must be run with the circuit conductors per NEC 250-57(b).

The concept of redundantly grounding such equipment is to ensure that a grounding means is always available in case of a ground-fault. This rule is aimed at protecting patients from electrical shock hazards.

MC/IG cable may be approved by the authority having jurisdiction (AHJ) to be installed in health care facilities and places of public assembly for the installation of IG circuits. MC/IG cable consist of

2 green grounding conductors, one with a yellow stripe to be identified as the isolated grounding conductor.

See Figure 6-7 for redundantly grounding equipment used to serve and care for patients per NEC 517-13(a), Ex. 1.

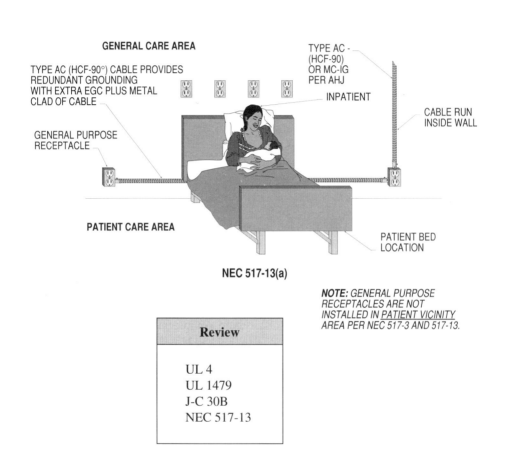

GENERAL CARE AREA

TYPE AC (HCF-90°) CABLE PROVIDES
REDUNDANT GROUNDING
WITH EXTRA EGC PLUS METAL
CLAD OF CABLE

TYPE AC -
(HCF-90)
OR MC-IG
PER AHJ

INPATIENT

CABLE RUN
INSIDE WALL

GENERAL PURPOSE
RECEPTACLE

PATIENT CARE AREA

PATIENT BED
LOCATION

NEC 517-13(a)

*NOTE: GENERAL PURPOSE
RECEPTACLES ARE NOT
INSTALLED IN PATIENT VICINITY
AREA PER NEC 517-3 AND 517-13.*

Review
UL 4
UL 1479
J-C 30B
NEC 517-13

Figure 6-7. AC Cable (HCF-90) may be used to wire-in receptacle outlets and fixed electrical equipment installed in patient care areas per the AHJ.

Special Facilities & Loc.

ISOLATION GROUNDING

Another application that could be served by HCF-90, the so-called "Hospital Grade" AC, or an MC cable with two grounding conductors is the isolated-ground application. In an effort to reduce the effects of electromagnetic interference on circuits supplying sensitive electronic equipment such as computers, peripherals, cash registers, test instruments, etc, a separate, additional grounding conductor (IG) that is isolated from the metal enclosure is provided in the cable. The isolated-ground is used to ground the sensitive equipment and the other grounding means is used to ground other enclosures of the circuit.

See Figure 6-8 for an application using armored cables to supply sensitive electronic equipment requiring an isolated ground in addition to a regular equipment grounding conductor per NEC 250-74, Ex. 4.

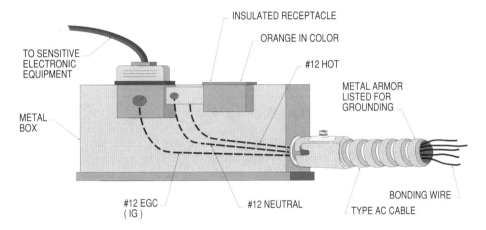

INSULATED RECEPTACLE

ORANGE IN COLOR

TO SENSITIVE ELECTRONIC EQUIPMENT

#12 HOT

METAL ARMOR LISTED FOR GROUNDING

METAL BOX

#12 EGC (IG)

#12 NEUTRAL

BONDING WIRE

TYPE AC CABLE

NEC 250-74, Ex. 4

Review
UL 4
UL 1479
NEC 333-19
NEC 333-20
NEC 250-74, Ex. 4
NEC 384-20, Ex.

Figure 6-8. AC Cable with armor, listed for grounding and equipped with an equipment grounding conductor, may be used to wire-in isolation receptacles.

HARMONIC CURRENTS

Another concern that has recently become a problem is overloaded neutral conductors on multiwire branch-circuits. This overload is the result of harmonic currents drawn by today's typical business equipment and electronic discharge lighting systems.

To address the inherent overload produced by many of today's loads, a special AC or MC cable is available with either an oversized neutral for each three/phase, four-wire circuit or a full-sized neutral (same size as the phase conductors) for each phase conductor in the cable. It is important to note that use of such cables do not eliminate the harmonic currents, but simply accommodates the additional neutral heating caused by the harmonic currents drawn by such loads.

See Figure 6-9 for a detailed application using armored cables to accommodate the heating effects caused by harmonic currents in neutral conductors per NEC 220-22 and Note 10(c) to ampacity Tables 0-2000 volts.

PLACES OF ASSEMBLY

A place of assembly is any single indoor space which is a portion of or a whole building that is designed or intended for use by 100 or more persons for assembly purposes. Such spaces include dining rooms, meeting rooms, entertainment areas (not stages, platforms, or projection booths), places of worship, bowling alleys, lecture halls, dance halls, etc. See NEC 518-2 for a list of spaces and buildings considered as places of assembly.

WIRING METHODS

MC cable can be used to wire-in Places of Assembly. However, AC cable is restricted to building areas that are not required by the local building codes to be fire-rated construction.

For example: AC cable may be used to wire-in spaces which are parts of buildings in which places of assembly exist but are portions that are not considered fire-rated construction.

See Figure 6-10 for wiring places of assembly with armored cables per NEC 518-4.

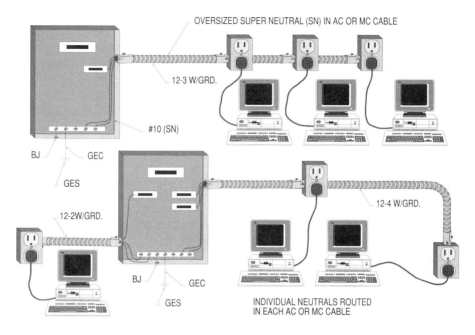

OVERSIZED SUPER NEUTRAL (SN) IN AC OR MC CABLE

12-3 W/GRD.

#10 (SN)

BJ

GEC

GES

12-2W/GRD.

BJ

GEC

GES

12-4 W/GRD.

INDIVIDUAL NEUTRALS ROUTED
IN EACH AC OR MC CABLE

NEC 220-22

Review
UL 1569 NEC 250-74, Ex. 4 NEC 384-20, Ex. NEC 220-22 Table 310-16, Note 10

Figure 6-9. AC and MC Cables are available with an extra or oversized neutral plus an equipment grounding conductor, to supply (nearly as possible) clean and noise free power to sensitive electronic equipment.

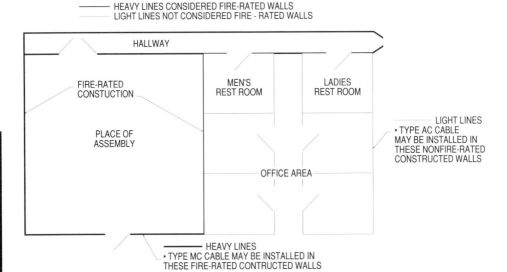

HEAVY LINES CONSIDERED FIRE-RATED WALLS
LIGHT LINES NOT CONSIDERED FIRE - RATED WALLS

HALLWAY

FIRE-RATED
CONSTUCTION

MEN'S
REST ROOM

LADIES
REST ROOM

LIGHT LINES
• TYPE AC CABLE
MAY BE INSTALLED IN
THESE NONFIRE-RATED
CONSTRUCTED WALLS

PLACE OF
ASSEMBLY

OFFICE AREA

HEAVY LINES
• TYPE MC CABLE MAY BE INSTALLED IN
THESE FIRE-RATED CONTRUCTED WALLS

NEC 518-4, Ex. 1

Type MC Cable may be used in these commercial buildings:	
Assembly Halls	Bowling Lanes
Exhibition Halls	Pool Rooms
Armories	Club Rooms
Dining Facilities	Places of Awaiting Transportation
Restaurants	Court Rooms
Church Chapels	Conference Rooms
Dance Halls	Auditoriums
Mortuary Chapels	Auditoriums within:
Museums	• Schools
Skating Rinks	• Mercantile Establishments
Gymnasiums	• Business Establishments
Multipurpose Rooms	

Figure 6-10. MC Cable may be used as a wiring method in places of assembly. However, AC Cable may only be used in locations considered by the local building code as nonfire-rated construction areas.

THEATERS, AUDIENCE AREAS OF MOTION PIC-TURE, TELEVISION STUDIOS AND SIMILAR LO-CATIONS

Such areas and locations pertain only to parts of a building that are utilized as a theater or similar location. Special wiring methods and techniques must be applied to these parts but not necessarily to the entire building.

For example: An auditorium in a school building used for performance and similar activities would be considered a place of assembly.

WIRING METHODS

MC cable can be used to wire any location in a place of assembly such as theaters, TV studios, etc. MC cable may be used to wire-in the auditorium and such areas as the stage, dressing rooms, and main corridors leading to the auditorium. However, both AC and MC cable may be used to wire-in other parts of the building that do not pertain to the use of the auditorium, which is used for performances or entertainment. Ex. 3 to NEC 520-4 permits AC cable to be installed in those portions of a building, which are not required to be fire-rated construction by local building codes.

See Figure 6-11 for a detailed application of wiring theaters, audience areas of motion picture and television studios, and similar locations per NEC 518-4.

MOBILE HOME AND MOBILE HOME PARKS

The elements that make up the service equipment and feeder-circuits to supply power to mobile home and mobile home parks are usually installed by contractors and electricians. Certain types of wiring methods are required to enclose and protect such elements per Article 550 of the NEC.

WIRING METHODS

MC cable with an overall PVC jacket for added mechanical protection may be directly buried in the ground and routed as a feeder-circuit between the service equipment and mobile home panelboard. Such cable can be used as service entrance conductors to supply power to the service equipment mounted on a pole or installed in a pedestal. The run of cable between the service equipment and mobile home panelboard must have four insulated conductors, which one is used as an equipment grounding conductor.

91

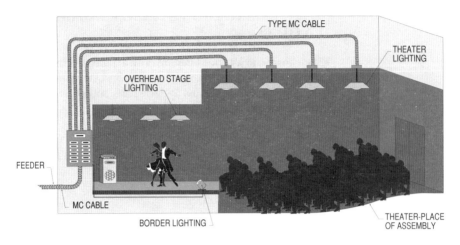

NEC 518-4

Review
UL 1569
NEC 518-2
NEC 518-4

Figure 6-11. MC Cable may be used to wire the auditorium (not including the stage) of theaters.

MC cable with a PVC jacket provides the flexibility needed to make the most difficult wiring job, such as building a service pole or routing a feeder-circuit to the mobile home panelboard, fast and easy. All the safety advantages of a metal-encased wiring system are provided with the flexibility and lower cost of installing a cable system.

By installing such cable systems, wasted time and effort on installing conduit and pulling conductors is eliminated. The man-hours between installing cable systems over conduit and conductors is greatly reduced.

See Figure 6-12 for a detailed application of wiring services and feeders supplying power for mobile home and mobile parks per NEC.

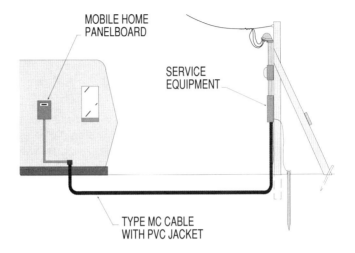

MOBILE HOME
PANELBOARD

SERVICE
EQUIPMENT

TYPE MC CABLE
WITH PVC JACKET

Special Facilities & Loc.

Review
UL 1569
NEC 550-11
NEC 550-23
NEC 550-24

Figure 6-12. MC Cable, with an overall outer PVC jacket, may be used as a feeder-circuit to supply power to a mobile home panelboard.

RECREATIONAL VEHICLES AND RV PARKS

In recreational parks, RV's must have a power supply provided by the park to which the RV connects its power supply assembly. RV parks usually provide this power with 15 to 50 amp receptacle outlets that are mounted on a power pole or pedestal. Outdoor wiring methods must be used for all the elements required to furnish safe and reliable power systems.

WIRING METHODS

MC cable with a PVC jacket can be utilized to wire-in the elements of service equipment mounted on the pole or in a pedestal. Such outdoor (approved) cables can be used to wire the receptacle outlets mounted below the service panel enclosure which are used to cord-and-plug connect the power supply assembly of the RV.

See Figure 6-13 for a detailed application of using MC cable with an overall PVC jacket to wire-in the elements required to supply power to RV's parked in sites of recreational RV parks per NEC.

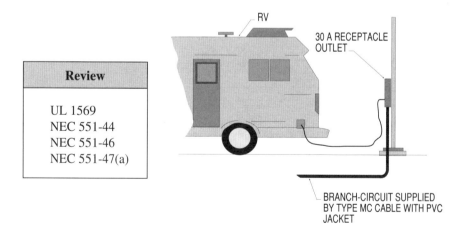

Review

UL 1569
NEC 551-44
NEC 551-46
NEC 551-47(a)

RV

30 A RECEPTACLE OUTLET

BRANCH-CIRCUIT SUPPLIED BY TYPE MC CABLE WITH PVC JACKET

Figure 6-13. MC Cable with an overall outer PVC jacket, may be used to wire-in receptacle outlets which are used to cord-and-plug connected RV's in recreational parks.

Special Facilities & Loc.

SWIMMING POOLS, FOUNTAINS, AND SIMILAR LOCATIONS

MC cable with an overall PVC jacket is suitable for use as a continuous run from the interior panel or subpanel of a residential or commercial building to outdoor or underground installations including swimming pool associated motors, filter pumps, or motorized automatic pool covers.

WIRING METHODS

Note: Not presently approved to be used to wire the wet-niche or dry-niche fixture for a swimming pool.

PVC jacket Type MC cable is listed to be installed outdoors in wet locations or directly buried in the earth or concrete. It is an excellent source for branch and feeder-circuits, plus power, lighting, control, and signal circuits. If used with specifically designed connectors, the cable's armor and grounding conductor combined provides an assured grounding path, which is particularly important in direct burial applications. **(See Note)**

See Figure 6-14 for an illustrated example of using MC cable with a PVC jacket to supply power to a swimming pool circulating pump motor.

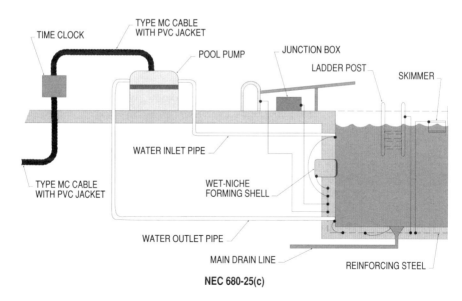

NEC 680-25(c)

Figure 6-14. MC Cable, with an outer overall PVC jacket, may be used as a branch circuit to wire in circulating pumps for pools and associated equipment.

	Chapter Six
Answers	**Questions**
T F	**1.** Type AC or MC cable can be installed at a savings up to 40% to 50% over EMT.
T F	**2.** Type AC or MC cable are not permitted to be utilized as wiring methods for other spaces used for environmental air.
T F	**3.** Red fire alarm/control cable is not suitable for use in ducts, plenums, and other spaces used for environmental air.
T F	**4.** Type MC cable is permitted as wiring method in locations containing flammable liquids and flammable gases.
T F	**5.** Type AC cable is not permitted for use in Class I, Division 2 hazardous locations.
T F	**6.** MC cable may be installed in cable trays if installed in such a manner to avoid tensile stress at termination points.
T F	**7.** Type MC cable is permitted to wire-in all types of electrical equipment installed in Class III, Division 1 and 2 locations.
T F	**8.** Type AC cable is not available with up to four conductors which includes an equipment grounding conductor.
T F	**9.** Health care facility cables are commonly referred to as HFC-90 or "Hospital grade" Type AC.
T F	**10.** A special Type AC or MC cable is available with either an oversized neutral for each three/phase, four-wire circuit or a full-sized neutral (same size as phase conductors) for each phase conductor in the cable.
_____	**11.** Places of assembly are any single indoor spaces which are parts of or a whole building that are designed or intended for use by _____ or more persons for assembly purposes.
_____	**12.** Type AC cable is restricted to building areas that are not required by local building codes to be _____ construction in places of assembly.

13. An _____ in a school building used for performance and similar activities would be considered a place of assembly.

14. Type MC cable with an overall PVC _____ for added individual protection may be directly buried in the ground and routed as a feeder-circuit between the service equipment and mobile home panelboard.

15. Type MC cable with a PVC jacket provides for _____ needed to make the most difficult wiring job of building a service pole or routing a feeder-circuit to the mobile home panelboard.

16. RV parks are usually provided with receptacle outlets in amperages of:

a. 15 to 40 amp
b. 20 to 40 amp
c. 15 to 50 amp
d. 20 to 50 amp

17. Type AC or MC cable can be installed at a savings over conductors up to:

a. 10 - 20%
b. 30 - 40%
c. 40 - 50%
d. 50 - 60%

18. Type MC cable can only be used in hazardous locations containing:

a. flammable liquids
b. containing dusts
c. containing flyings
d. all of the above

Answers	Questions
_____	**19.** Type MC cable used as a wiring method in Class I, Division 2 locations, may only be installed in which cable trays: **a.** ladder-type **b.** ventilated trough **c.** ventilated channel **d.** all of the above
_____	**20.** Which of the following are not considered a place of assembly: **a.** restaurant **b.** dwelling unit **c.** assembly hall **d.** court room

CHAPTER 7

OLD WORK

One of the most common problems encountered by designers and install-ers in remodeling is the routing of the new branch-circuits. A very common problem is fishing wiring methods down enclosed walls, ceil-ings, floors, etc.

AC and MC cables are excellent for fishing and wiring existing premises due to their flexibility. They can be installed in a one-step process which simplifies cable re-routing for equipment relocation due to renovation and rehab projects.

SELECTING BOXES

Before actually routing the armored cable, the right boxes must be selected and holes cut in the walls or ceiling coverings to install them. Electrical boxes for rewiring are available in many types and sizes. Such boxes are sometimes referred to as "old work" or "cut-in" boxes and they differ from many new work boxes in that they mount easily where wall materials are already in place. Boxes must be selected large enough to hold all the necessary conductors and number of lighting fixtures, switches, receptacles or combinations.

See Figure 7-1 for the various boxes available for use in wiring existing buildings.

CHECKING FOR OBSTRUCTIONS

Hunting the location of studs may be an uncertain task. In old construction, they may be found on 16 in. centers but not always. Sometimes studs can be found by knocking on the wall with your knuckles and listening for a solid rap. A more accurate method is to use a stud finder which is equipped with a magnetic pointer that is attracted to the metal nail heads.

The surest procedure is to drill a small test hole where the box is to be mounted and bend an eight to ten inch length of stiff wire and push it through the hole and rotate it. If it hits a stud while rotated right or left, move over until an empty space is found.

See Figure 7-2 for methods of locating studs with wall coverings already finished.

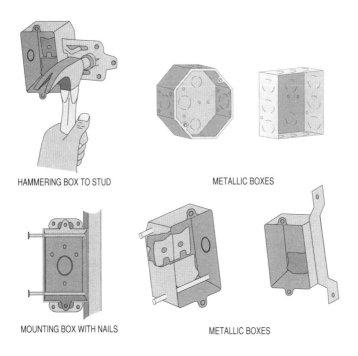

HAMMERING BOX TO STUD

METALLIC BOXES

MOUNTING BOX WITH NAILS

METALLIC BOXES

Figure 7-1. Installers must select the type of box needed to support devices, etc. There are a variety of boxes available for such use.

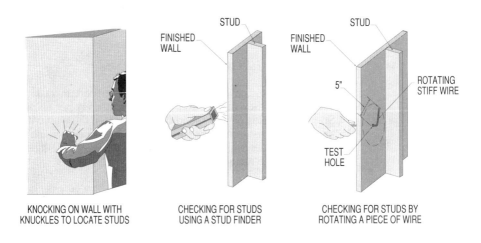

KNOCKING ON WALL WITH
KNUCKLES TO LOCATE STUDS

CHECKING FOR STUDS
USING A STUD FINDER

CHECKING FOR STUDS BY
ROTATING A PIECE OF WIRE

Figure 7-2. Locating obstructions behind finished walls.

MARKING HOLES FOR BOXES

Once a clear space has been found, mark the wall or ceiling for cutting the box hole. For a plain wall-mounted box, place box face as a template to trace the box outline. Omit the adjustable top and bottom ears that are used to support the box when installed.

See Figure 7-3 for marking walls and ceilings to cut box holes.

CUTTING HOLES FOR BOXES

After the holes are marked, the next step is to cut them. The most common way to cut holes is to punch a tiny hole and use a saber saw or a keyhole saw to cut the hole. After cutting the hole, fish in the fish tape and pull AC or MC cable to the outlet, equipment, or fixture served.

See Figure 7-4 for methods used to cut holes to mount boxes in walls with finished coverings.

MOUNTING BOXES

After the holes have been cut and the armored cable is routed from the power source to the new box location, the remaining job is to mount the box and connect the fished-in cable.

The AC or MC cable is connected by use of a cable clamp or slipping a connector on the cable end and fastening it to the box. Leave at least six inches of conductors sticking out of the box for connection of devices, fixtures, etc.

See Figure 7-5 for the type of boxes used to mount in walls that are finished with coverings.

Old Work

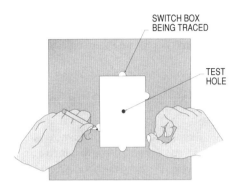

Figure 7-3. Using a switch box to trace and mark box hole location.

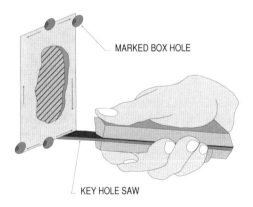

Figure 7-4. Cutting box holes using a keyhole saw.

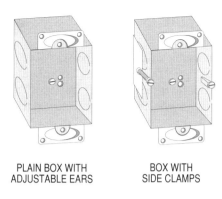

PLAIN BOX WITH
ADJUSTABLE EARS

BOX WITH
SIDE CLAMPS

Figure 7-5. Electrical boxes for use in existing finished walls are available in many types and sizes.

BOXES WITH EARS

Once the box is placed in the hole, if necessary, adjust the ears so the front edge of the box is flush with finished wall material. The box ears are used with hold fast bars to secure the box. Needle nose pliers are used to bend the tops of the hold fast bars to the sides of the box.

See Figure 7-6 for installing boxes with ears and securing the tops of hold fast bars.

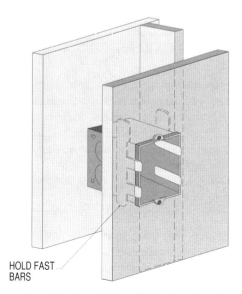

HOLD FAST
BARS

Figure 7-6. Mounting a box in existing finished wall using hold fast bars.

METALLIC CUT-IN BOXES

The mounting of this type of box is a one time installation, for once inside the wall or ceiling, the teeth flare away from such boxes, which makes it very difficult to remove. The teeth are pushed away from the box by tightening the screw at the back of the box.

See Figure 7-7 for directions on how to mount metallic cut-in boxes.

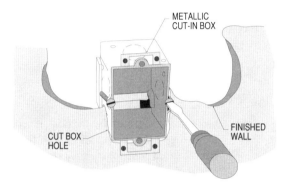

Figure 7-7. Mounting a box in existing finished wall using a metallic cut-in box.

FISHING CABLES

Before attempting to fish AC/MC cables, obtain the proper fishing gear. For long cables runs, it is recommended to use fish tapes and for shorter runs use a straightened coat hanger or piece of wire which is bent into a tight hook. When fishing from above and down to a box cut out in the finished wall, a small fixture chain can be used to pull-in the cable.

IN WALLS

Installers have a big problem fishing cables inside finished walls. The enclosed walls may have headers installed between the framing members. When this situation occurs the headers have to be drilled with an extension drill bit or the wall material covering the framed studs must be removed and the header notched or taken out.

Installers usually fish down between the framing members with a fixture chain because the chain has flexibility and is easy to fish.

The next step is to tie the AC/MC cables to the chain and tape over tight and smooth, so the cable can be pulled easily through the holes. Fish tape may also be used to fish cable.

See Figure 7-8 for the procedure used to fish armored cable inside a finished wall from an attic to a box outlet.

Note: Metal studs are used for commercial wiring. Wood studs are used in Figure 7-8 to illustrate NEC 300-4.

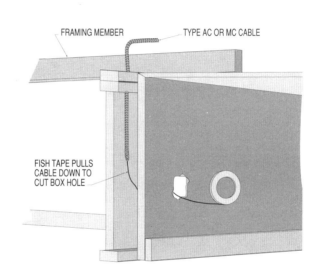

FRAMING MEMBER   TYPE AC OR MC CABLE

FISH TAPE PULLS
CABLE DOWN TO
CUT BOX HOLE

Figure 7-8. Fishing armored cable in existing finished wall.

IN CEILINGS

AC or MC cable can be fished in an inaccessible attic from a cut fixture hole to a cut ceiling hole and top wall hole to a cut hole for a wall switch. After holes are cut, run fish tape from the top wall hole to the cut switch box hole and connect to switch loop cable and pull it up and out of the hole. Next run fish tape from fixture hole to top ceiling hole and pull loop switch cable to and out of cut fixture hole. **See Figure 7-9** for a detailed illustration of how to fish cables in attics that are not accessible.

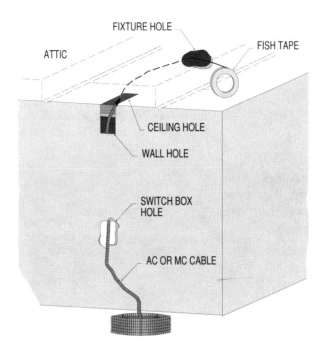

Figure 7-9. Fishing armored cable in existing finished attics with no access.

IN FLOORS

When fishing cables in enclosed covered walls, drill a small guide hole through the floor beneath your cut box opening. Using a drill bit large enough to drill a passage hole for the armored cable, drill the passage hole next to the guide hole. Attach cable to a fish tape and pull the cable from the finished basement or from underneath the pier and beam floor.

See Figure 7-10 for a detailed illustration of how to drill and pull armored cable from basements and underneath accessible floors.

BEHIND BASEBOARDS

Note: The power supply cable is fished from overhead or from underneath the floor.

Cut holes for boxes and remove existing baseboard between box locations. Next, drill hole through finished wall below each box and cut access channel in wall to connect holes for boxes. The final job is to fish the armored cable down through one box hole and along the cut channel where it is fished up to the other box. **(See Note)**

See Figure 7-11 for installing AC or MC cable behind baseboards in existing buildings.

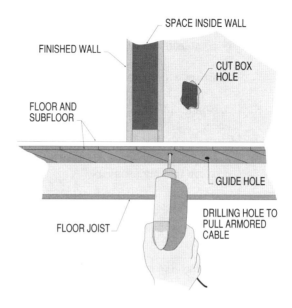

Figure 7-10. Marking guide hole and drilling passage holes to pull armored cable from space underneath finished floor.

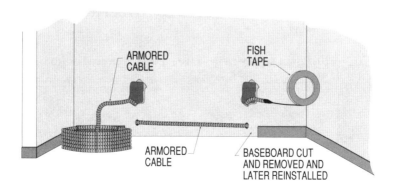

Figure 7-11. Fishing armored cable behind finished baseboard in existing covered walls.

Chapter Seven	
Answers	**Questions**
T F	**1.** Boxes must be selected large enough to hold all the necessary conductors and number of lighting fixtures, switches, receptacles or combinations.
T F	**2.** Boxes with ears are used with hold fast bars to secure the box.
T F	**3.** Nonmetallic cut-in boxes are allowed to be used more than once.
T F	**4.** A fixture chain is usually used by installers to fish down between framing members.
T F	**5.** Type AC or MC cable can not be fished in an inaccessible attic.
_____	**6.** A _____ pointer is attracted to the metal wall heads when locating a stud.
_____	**7.** Once a clear space has been found, _____ the wall or ceiling for cutting the hole.
_____	**8.** The most common way to cut holes is to punch a tiny hole and use a saber saw or a _____ saw to cut the hole.
_____	**9.** When fishing cables in enclosed covered walls, drill a small _____ hole through the floor beneath your cut box opening.
_____	**10.** Power supply cable is fished from overhead or from _____ the floor.

REVIEW QUESTIONS - ANSWER SHEET

CHAPTER 1
GENERAL REQUIREMENTS

1. T
2. T
3. F
4. T
5. F
6. T
7. T
8. T
9. F
10. T
11. interlocking
12. copper
13. equipment
14. listed
15. ground
16. not
17. impervious
18. varnished
19. embedded
20. bushing
21. d - 14
22. a - 1
23. c - 16
24. d - green
25. b - thermoplastic

CHAPTER 2
CONDUCTOR AMPACITY

1. T
2. F
3. T
4. T
5. F
6. T
7. F
8. T
9. F

10. F
11. three
12. higher
13. Note 8(a)
14. 51%
15. bonding
16. c - 86°F
17. b - 82%
18. c - 80%
19. a - 22.96 A
20. a - 10 AWG

CHAPTER 3
OVERCURRENT PROTECTION

1. T
2. F
3. F
4. T
5. T
6. F
7. F
8. T
9. T
10. T
11. lower
12. 440
13. unknown
14. markings
15. 100%
16. b - 18.1 A
17. a - 23.5 A
18. a - 60°C
19. b - 75°C
20. c - 125%

CHAPTER 4
INSTALLATION

1. T
2. F

3. F
4. F
5. T
6. T
7. F
8. T
9. T
10. F
11. T
12. T
13. F
14. F
15. T
16. 6 1/2 ft.
17. 5 1/2 ft.
18. 12 in.
19. 1/16 in.
20. 60°
21. d - 6 ft.
22. c - 7 ft.
23. d - 6 ft.
24. b - 2 ft.
25. a - 6 ft.

CHAPTER 5
SPECIAL CIRCUITS AND
EQUIPMENT

1. T
2. T
3. F
4. T
5. F
6. F
7. T
8. T
9. F
10. F
11. branch-circuits
12. 150%, 200%
13. No. 18, No. 12
14. security
15. intrinsically
16. a - No. 8
17. d - all of the above

18. a - No. 22 to No. 24
19. c - Class II, Division 2
20. c - Class III, Division 1 and 2

CHAPTER 6
SPECIAL FACILITIES AND
LOCATIONS

1. T
2. F
3. F
4. T
5. T
6. T
7. T
8. F
9. T
10. T
11. 100
12. fire-rated
13. auditorium
14. jacket
15. flexibility
16. C
17. B
18. D
19. D
20. B

CHAPTER 7
OLD WORK

1. T
2. T
3. F
4. T
5. F
6. magnetic
7. mark
8. keyhole
9. guide
10. underneath

THE 20 MOST ASKED QUESTIONS ABOUT
ARMORED CABLE

1. Where can armored cable be used?

Generally, Section 333-3 of the National Electrical Code covers the permitted uses, and Sections 300-22(c), 410-67(b), 517-13, 517-81, 604-6(a), 620-21, 645-5(d)(2) and others covers specific applications.

2. Does the NEC contain any requirements for termination of the bond wire in armored cable?

No. It may simply be cut-off at the end of the armor. However, many electricians wrap the bond wire over the insulating bushing and around the armor to secure the bushing in place.

3. Can AC/MC cables be used in the return air space above hung ceilings?

Yes. NEC Section 300-22(c) defines this space used for environmental air and Type AC/MC cables are allowed in this location.

4. May armored cable be used without a separate equipment grounding conductor?

Yes. The armor and bond wire combination is recognized as an acceptable equipment grounding conductor in accordance with Section 250-91(b) of the NEC.

5. Is armored cable with 90°C conductor insulation available to meet the requirements of the Exception to Section 333-20 of the 1993 Edition of the NEC?

Yes. armored cable with 90°C conductor insulation is available. A tag on the coil or reel identifies the cable as "Type ACTHH".

6. May armored cable be used as a "fixture whip" to supply lay-in fluorescent fixtures?

Yes. Section 410-14 and 410-67(b) permit running armored cable branch-circuit conductors directly to a lighting fixture. Section 333-7, Exception 3 in the 1993 NEC revises the support requirements for armored cable. It allows the last 6 feet to be installed unsupported as a whip for connection to lighting fixtures and equipment in accessible ceilings.

7. How can I tell the difference between armored cable and metal-clad cable, since they both have the same outside appearance?

Armored cable will always have an aluminum bond wire under the armor, and the individual insulated conductors will each have a kraft paper wrap. Metal-clad cable does not contain either of these items.

8. Can armored cable be installed under raised floors in data processing rooms?

Yes. According to NEC Section 645-5(d)(2) armored cable can be used to supply branch-circuit receptacles in this location.

9. What is the maximum number of conductors and the sizes available for armored cable?

According to UL Standard 4 for armored cable, a maximum of 4 circuit conductors with or without an equipment ground is allowed. The copper size range is from No. 14 AWG to No. 1 AWG.

10. Is armored cable an acceptable wiring method for patient care areas in a hospital?

Yes. Section 517-13 recognizes the use of Type AC with a green insulated copper grounding conductor for this application.

11. What types of fittings must be used with armored cable?

Listed and labeled armored cable fittings have inspection slots so that the insulating bushing is visible for inspection. These fittings are supplied in containers specifically identified for this application. Metallic boxes which have built-in clamps for armored cable are also available.

12. Can I wire a dwelling unit with armored cable?

Yes. This wiring method is excellent for dwelling units since the metallic armor does not support combustion, is rodent proof, and also offers resistance to nail penetration and crushing.

13. Where 4-conductor armored cable is used for a three/phase, 4-wire branch-circuit to supply fluorescent fixtures, do conductor ampacity derating factors apply?

Yes. Note 10(c) to Tables 310-16 thru 310-19 indicates that the neutral conductor is considered a current-carrying conductor under these conditions. Note 8(a) requires that the ampacities of conductors listed in Table 310-16 be reduced to 80 percent of these values. Where Type ACTHH cable is used and the cable is not installed in thermal insulation, the derating factor is applied to the ampacities under the 90°C column.

14. What electrical tests are performed at the factory on AC/MC cables?

All finished coils and reels of armored cable are 100 percent tested with both a dielectric withstand test of the insulation and a continuity test to ensure the integrity of the product.

15. Does connecting the bond wire in a No. 14-2 armored cable to the grounding terminal of a 15 ampere, duplex receptacle satisfy the requirements of the NEC?

No. It is not necessary to connect the bond wire to the grounding terminal. The armor and bond wire combination connected to the junction box satisfies the requirement of Section 250-91 for an equipment grounding conductor. However, a bonding jumper may be required between the receptacle and the box to comply with Section 250-74.

16. May armored cable be used in high-rise office building?

Yes. The height of a building does not have any bearing on the use of this wiring method.

17. Must insulating bushings be used at all armored cable terminations?

The NEC requires insulated bushings to be inserted between the conductors and the armor at every termination. (See Section 333-9)

18. What does "cable tray rated" mean as applied to armored cable?

This indicates that the cable has passed the 70,000 BTU vertical tray flame test as defined in UL Standard 4 for armored cable, and that the cable may be installed in cable trays.

19. Does all armored cable require a bond wire? What is the material and size of the bond?

Yes. All armored cable, other than lead sheathed cable (Type ACL), must have an uninsulated bond wire throughout its entire length. The bond wire must be aluminum and cannot be smaller than No. 16 AWG.

20. Why is new armored cable bright and shiny, whereas older cable is dull gray?

This is a normal aging process for galvanized steel and does not affect the equipment grounding effectiveness of the armor/bond wire combination.

A COMPARISON OF AC AND MC CABLES

ARMORED CABLE - TYPE AC	METAL CLAD CABLE - TYPE MC
NEC Article 333 - UL 4	NEC Article 334 - UL 1569
Branch-circuits and feeder cables exposed or concealed work.	Services, feeders, and branch-circuits, power, lighting and control circuits, exposed or concealed work.
Sizes No. 14 through No. 1 AWG.	Sizes No. 18 AWG through No. 250 KCMIL (equipment limit).
Factory assembled 2, 3, or 4 conductors, with or without an insulated or uninsulated grounding conductor.	1 or more conductors, factory assembled with an insulated grounding conductor (bare grounding conductor also available).
90°C rated, 0-600 volts.	90°C rated, 600 volts (75°C wet locations).
Dry locations only.	Wet or dry locations - indoors or outdoors, aerial installations, direct buried.
Equipment bonding conductor - No. 16 AWG alloy bond wire in intimate contact and combined with the armor. Separate grounding conductor optional.	Separate equipment grounding conductor cabled with phase and neutral conductors. No bond wire required.
Paper wrap required on individual conductors. Overall binder of assembly tape not required.	Over binder (assembly tape) required. No paper wrap on individual conductors. Fillers and jackets over and under the armor may be required depending upon installation and application.
Minimum bending radius = 5 times cable O.D.	Minimum bending radius = 7 times cable O.D. (interlocked armor type).
Ampacities NEC 310-15 and Table 310-16.	Ampacities NEC 310-15 and Table 310-16.
Armor: Steel, galvanized on all sides or aluminum alloy.	Galvanized (all 4 sides) steel or aluminum alloy armor.

A COMPARISON OF AC AND MC CABLES

ARMORED CABLE - TYPE AC	METAL CLAD CABLE - TYPE MC
Anti-short bushings required.	Anti-short bushings optional.
Marking: Indent armor required.	Marker tape under armor. Print legend on jacketed cable.
Cable tray use - Yes, if listed with UL (passes Vertical Flame Test).	Cable tray use - Yes, if listed with UL (passes Vertical Flame Test).
Uses permitted: For branch-circuits and feeders in both exposed and concealed work, dry locations for under plaster extensions, embedded in plaster finish on brick or masonry (except damp or wet locations), fished or run in air voids of masonry block or tile walls.	Uses permitted: For services, feeders and branch-circuits, for power, lighting control and signal circuits, indoors or outdoors, exposed or concealed, direct buried when identified for such use, cable tray, in any approved raceway, as open runs of cable, as aerial cable on a messenger, in hazardous conditions as permitted in Articles 501, 502, and 503, in dry locations or wet locations when identified for such use.

MC CABLE vs. HCF - 90

CONSTRUCTION:	MC	MC/IG	HCF - 90
Conductors	Copper	Copper	Copper
Insulation	THHN	THHN	THHN
Cable Tape	Dielectric mylar overall	Dielectric mylar overall	Paper wrap on each
Grounding and bonding	1 grounding system; Armor and green ground	2 grounding systems; Armor and plain bond wire (equipment) Insulated copper wire (isolated)	2 grounding systems; Armor and bond wire (equipment) Insulated copper wire (isolated)
Armor	Galvanized steel or aluminum	Galvanized steel or aluminum	Galvanized steel or aluminum
Bushings	Optional	Optional	Required
Connectors	Must be MC - No direct bearing on set screws	Must be MC - No direct bearing on set screws	Must be AC - Direct bearing on screws for steel armor only

APPLICATIONS:

MC	MC/IG	HCF - 90
Places of public assembly over 100 Theaters Motion picture and TV studios Article 334 and 518 of the NEC	Computer power circuits Isolated ground circuits Articles 334 and 645 of the NEC	Hospitals, nursing homes, and medical centers Computer power circuits Isolated ground circuits Articles 333, 517, and 645 of the NEC

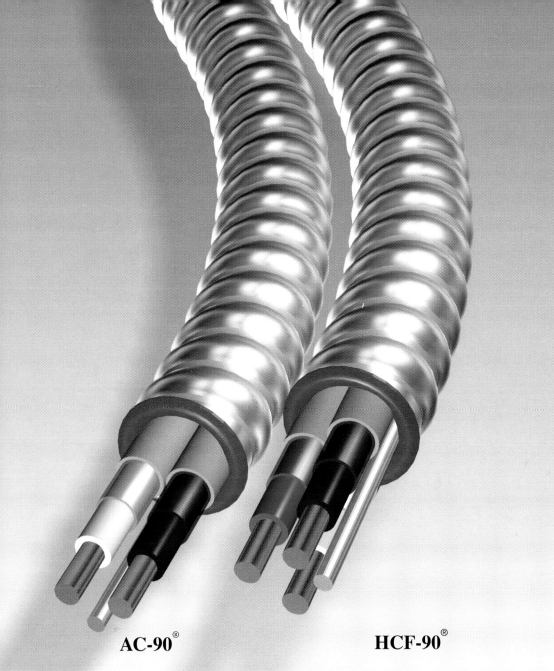

AC-90®

For continuous duty in higher ambient temperature conditions, you need AC-90 Armored Cable from AFC. It's the one designed for full 24 hour service at elevated temperatures, up to 90°C, in commercial, multi-family residential, industrial structures, and other structures.

HCF-90®

Inside health care facilities, arteries of electrical conduit carry the power that keeps everything operating smoothly. Because safety is so critical, the National Electrical Code and other standards require specific safeguards like the use of an insulated green copper grounding conductor which is factory installed in HCF-90 Health Care Facilities Cable.

AC - 90®

More useable power

Designed for higher thermal capability, AC-90 Armored Cable withstands greater heat, delivers full rated ampacity after derating for elevated ambient temperatures and, thereby provides more useable power per conductor size. AC-90 is made with a strong galvanized steel armor.

Result? You get more useable current with AC-90 and provide for future increases in the use of electricity in a variety of applications such as higher temperature lighting fixtures (marked for up to 90°C rated wire), and in spaces above suspended ceilings used for environmental air (see NEC Section 300-22(c).

90°C rated conductors make the difference

A product of high temperature wiring technology, AC-90 Armored Cable is specifically suited for use in heated environments including structures located in hot climates, in multi-unit dwellings located anywhere, near furnaces, in high-temperature machine shops, in foundries, adjacent to baking or industrial process ovens... wherever higher ambient temperature conditions exist.

Energy efficient

AC-90 Armored Cable also has all the reliability for higher voltage lighting systems such as the 480Y/277V energy systems used in the construction/restoration/rehabilitation of commercial, residential or industrial buildings. AC-90 is ideally suited for the modern 480Y/277V energy systems, which require less ballasting and use smaller wires automatically adding to the total efficiency of the system.

HCF - 90®

AFC HEALTH CARE FACILITIES CABLE — HCF-90

In hospitals, nursing homes, and other health care facilities, where lives depend on the proper functioning of electrical systems, only the finest wiring systems will do. HCF-90 is made with a strong galvanized steel armor.

Inside health care facilities, arteries of electrical conduit carry the power that keeps everything operating smoothly. Because safety is so critical, the National Electrical Code and other standards require specific safeguards like the use of an insulated green copper grounding conductor which is factory installed in HCF-90 Health Care Facilities Cable.

AFC's HCF-90 (Health Care Facilities Cable) meets the most stringent industry requirements. That's why contractors who recognize quality choose to specify AFC.

FULLY APPROVED AND CERTIFIED

When you order AFC's HCF-90 Health Care Facilities Cable, you are assured of compliance with NEC Article 517 and Federal Specification J-C-30B. HCF-90 is UL listed and labeled by Underwriters Laboratories as armored cable Type ACTHH per Standard UL 4 (Section 517-13 indicates that in patient care areas, an insulated copper grounding conductor must be used and UL Standard 4 specifies that is must be green.)

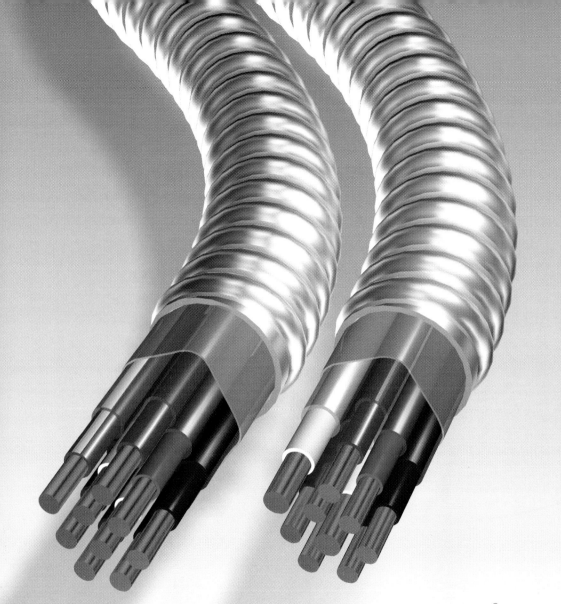

Home Run® Cable

Home Run Cable from AFC is the fastest way to connect all your power, lighting, control, and signal circuits. This unique type of MC Cable is designed to hold a multiconductor assembly inside a galvanized steel interlock armor. Its flexibility and convenience will save you money because Home Run Cable eliminates wasted time and effort on the job and greatly reduces man-hours during installation.

Super Neutral Cable®

Super Neutral Cable is a metal-clad, Type MC Cable, manufactured with an oversized neutral conductor or one neutral per phase for three-phase / four-wire power supply systems to computers (with dc drive fan motors, tape, and disk drives), office machines, programmable controllers, and similar electronic equipment where non-linear switching loads produce additive, odd order harmonic currents which may create overloaded neutral conductors.

HOME RUN CABLE®

Home Run Cable Defeats Pipe and Wire

You can save as much as 50% over pipe and wire when you install Home Run Cable. It installs quickly and easily - just like standard armored cable - so you save both time and money. A simple hand-held cutter is the only tool you'll need. Gone are the time-consuming, back-breaking, wirepull methods of EMT and wire. AFC has a more efficient wiring system designed to make every wiring job more profitable - for you.

Home Runs Are Our Business

AFC designed Home Run Cable to help you go the distance...with one cable. Perfect for multi-family residential, commercial, and industrial buildings. Home Run Cable gives you 6, 8, 12, or 16 insulated conductors inside a single inter-locked armor. It can be used exposed or con-cealed, in cable trays, in approved raceways, or as supported runs of cable. Home Run Cable gives you the flexibility needed to make your most difficult wiring jobs easy, safe, and less ex-pensive. Home Run Cable must be supported every 6' unless fished.

AFC Plays It Safe

Up to and including 600 volts can be carried by this quality cable from AFC. Home Run Cable's THHN insulated copper conductors provide a 90°C temperature rating for maximum safety. You get all the fire safety advantages of metal-en-cased wiring with the flexibility and cost advan-tages of a cable system.

SUPER NEUTRAL CABLE®

Description

Super Neutral Cable is a metal-clad, Type MC Cable, manufactured with an oversized neutral conductor or one neutral per phase for three-phase, four-wire power supply systems to com-puters (with dc drive fan motors, tape, and disk drives), office machines, programmable control-lers, and similar electronic equipment where non-linear switching loads produce additive, odd or-der harmonic currents which may create over-loaded neutral conductors.

The oversized neutral conductor(s) are sized 150% to 200% of the phase conductor ampacity to minimize the effects of harmonics generated by the non-linear loads. The neutral per phase (striped with color to match the phase conduc-tor) accomplishes the same objective.

Applications

Super Neutral Cable is listed and labeled by UL and is ideal in commercial, industrial, and util-ity applications where reliability is the major con-cern, maximum performance is demanded, space is limited, and ease of installation is critical. These MC Cable system components are installed in a one-step process so labor costs are greatly reduced. It can be routed much easier than rigid conduit which is a special advantage when several bends are required. It can be removed, rerouted, or completely relocated.

Super Neutral Cable can be used under computer room floors, raised floors, or overhead in the space above hung ceilings used for environmental air handling. It handles branch and feeder, plus power, lighting, signal, and control circuits in dry locations.

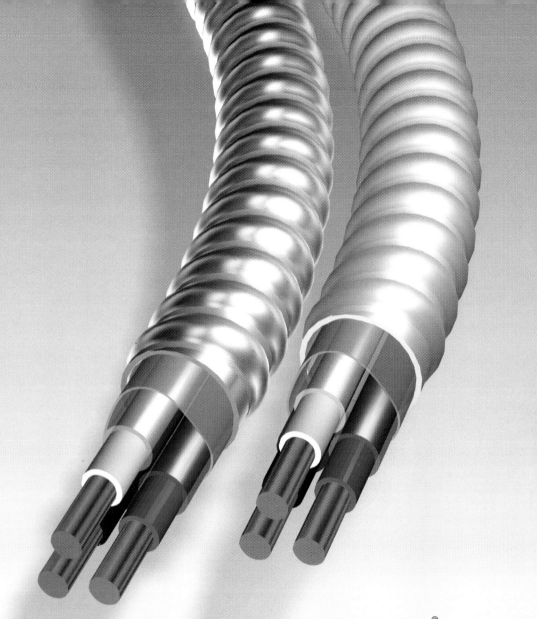

MC Cable

Gone are the time-consuming, lift-and-lug, bend-and-strain methods required for EMT and wire.

AFC was the first to develop a new metal-clad cable system which offers contractors a cost-efficient replacement for pipe-and -wire systems --especially when it comes to installing reliable power, lighting, control, and signal circuits in commercial buildings and large places of public assembly.

MC Lite®

MC Lite is ideal in commercial, industrial and utility applications where reliability is the major concern, maximum performance is demanded, space is limited, and ease of application is critical.

MC CABLE

MC LITE®

DESCRIPTION

Solid or stranded copper conductors; THHN (90°C) insulation; bare or insulated ground wire; all conductors cabled; marker-type inserted; cable tape overall and sheathed in a galvanized steel interlocked armor.

APPLICATIONS

Type MC cable is ideal in commercial, industrial and utility applications where reliability is the major concern, maximum performance is demanded, space is limited and ease of installation is critical.

This advanced cable system offers particular advantages over EMT and wire systems in power and lighting circuits installed in manufacturing and processing plants, as secondary feeders in industrial and commercial distribution systems, and for supplying power to station auxiliaries in power stations and substations.

Type MC may be used exposed or concealed, in cable trays, approved raceways or as an open run of cable. Type MC conforms to NEC Article 334 and is UL listed and labeled. (UL 1569)

DESCRIPTION

Solid or stranded copper conductors; THHN (90°C) insulation; bare or insulated ground wire; all conductors cabled; marker-type inserted; cable tape overall and sheathed in an aluminum interlocked armor.

This advanced cable system offers particular advantages over EMT for applications in power and lighting circuits installed in manufacturing and processing plants, as secondary feeders in industrial and commercial distribution systems, and for power supplies to station auxiliaries and substations.

APPLICATIONS

MC Lite may be used exposed or concealed in cable trays, approved raceways or as an open run of cable. MC Lite conforms to NEC Article 334 and is UL listed and labeled. (UL 1569)

MC Lite cable offers contractors a cost-efficient replacement for pipe-and-wire systems—especially when it comes to installing reliable power, lighting, control and signal circuits in commercial buildings and large places of public assembly.

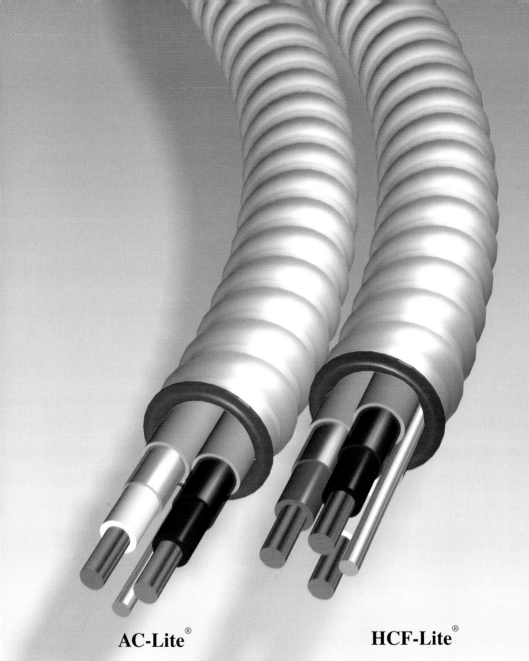

AC-Lite®

AC - Lite is ideal for commercial, industrial, and multi-family residential branch-circuit and feeder wiring up to 600 volts. It offers reliability for 120/208V systems and for higher voltage systems such as 480Y/277V. AC-Lite can be hardwired into 90°C rated fixtures and is ideal for fixture whip installations regardless of the fixture tail length.

HCF-Lite®

Inside health care facilities, arteries of electrical conduit carry the power that keeps everything operating smoothly. Because safety is so critical, the National Electrical Code and other standards require specific safeguards like the use of an insulated green copper grounding conductor which is factory installed in HCF- Lite Health Care Facilities Cable.

AC - LITE®

ALUMINUM ARMORED CABLE

AC-Lite is ideal for commercial, industrial, and multi-family residential branch-circuit and feeder wiring up to 600 volts. It offers reliability for 120/208V systems and for higher voltage systems such as 480Y/277V. AC-Lite can be hardwired into 90°C rated fixtures and is ideal for fixture whip installations regardless of the fixture tail length with support after six feet. AC Lite is made with a strong lightweight aluminum armor.

HIGH QUALITY

AFC's coiling methods practically eliminate armor imperfections plus each coil is subjected to additional high-voltage and continuity tests to insure trouble-free performance.

AFC uses special lubricants which evaporate quickly resulting in an extra clean product without the oily film that gathers dust and dirt in storage.

The cable is coiled to make stacking safe and easy. Wire ties are secured so the material is delivered in first class condition.

AVAILABILITY

AC-Lite is available in sizes #14 to 1 AWG with 2, 3, or 4 conductors. It is made with standard color coding (red, white, black, blue) but can also be made with special color (brown, orange, yellow & gray), upon request.

HCF - LITE®

HCF- LITE

In hospitals, nursing homes, and other health care facilities, where lives depend on the proper functioning of electrical systems, only the finest wiring systems will do. HCF-Lite is made with a strong lightweight aluminum armor.

Inside health care facilities, arteries of electrical conduit carry the power that keeps everything operating smoothly. Because safety is so critical, the National Electrical Code and other Standards require specific safeguards like the use of an insulated green copper grounding conductor which is factory installed in HCF-Lite Health Care Facilities Cable.

AFC Armored HCF-Lite Health Care Facilities Cable meets the most stringent requirements. That's why contractors who recognize quality choose to specify AFC.

FULLY APPROVED AND CERTIFIED

When you order AFC's HCF-Lite Health Care Facilities Cable, you are assured of compliance with NEC Article 517 and Federal Specification J-C-30B. It is UL listed and labeled by Underwriters laboratories as armored cable Type ACTHH per Standard UL 4 (Section 517-13 indicates that in patient care areas, an insulated copper grounding conductor must be used and UL Standard 4 specifies that is must be green.)

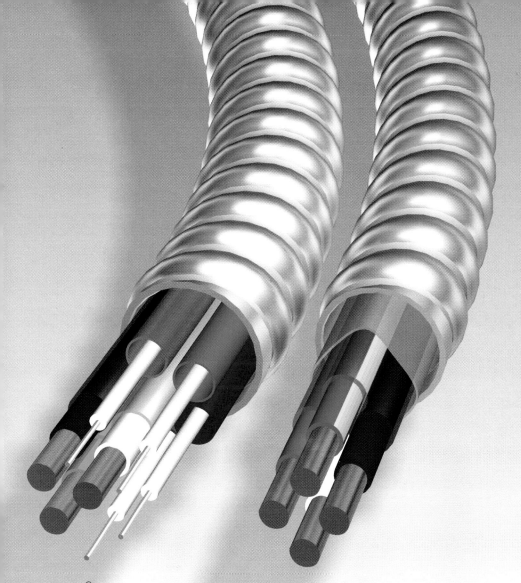

MC/OF® Power & Fiber Optic

Composite MC/OF Cable features efficiency of design, space, physical protection, and security. It is ideal for robotics and factory installations, video conferencing, and medical imaging applications in health care facilities. Custom cables with a variety of power conductors and optical fibers can be made to meet specific customer needs. UL listed and labeled in accordance with UL Standard 1569.

MC/IG® Isolated Ground Cable

MC/IG is the ideal cable for critical circuits where isolated or quiet grounding is desired.

The Isolated Ground Circuit can be made relatively free of electrical noise when installed through the intermediate panel boards without being connected to their equipment grounding terminal and grounded directly at the equipment grounding conductor terminal of the derived system, or service, in compliance with NEC Section 250-74, Ex. 4.

MC/OF® POWER & FIBER OPTIC

MC/IG®

MC/OF POWER & FIBER OPTIC

ISOLATED GROUND CABLE

The MC/OF Cable combines both current-carrying conductors and optical fibers in accordance with Section 770-4(c) and 770-52(a) of the 1993 NEC and UL Standard 1569.

MC/OF cable combines two installation steps into one. Both the premise wiring and power cables are installed in one cable, saving one half of the installation cost.

Composite MC/OF Cable features efficiency of design, space, physical protection, and security. It is ideal for robotics and factory installations, video conferencing, and medical imaging applications in health care facilities. Custom cables with a variety of power conductors and optical fibers can be made to meet specific customer needs.

AFC's MC/OF Cable offers cost savings of up to 50% over pipe and wire and may be used in cable trays, above hung ceilings in accordance with NEC 300-22(c), and under computer room floors.

The Isolated Ground Circuit can be made relatively free of electrical noise when installed through the intermediate panelboards without being connected to their equipment grounding terminal and grounded directly at the equipment grounding conductor terminal of the derived system, or service, in compliance with NEC Section 250-74, Ex. 4. MC/IG is ideal for this application.

A conventional grounding network can act as a giant antenna conducting noise for Electromagnetic Interference (EMI), adversely affecting the operation of electronic equipment, which act as transient signal "receivers".

MC/IG Cable is ideal for use with electronic computer/data processing equipment in commercial and industrial installations in accordance with NEC Article 645 and is two hour fire rated per UL Fire Wall Penetration Test (UL 1479), for walls, ceilings, and floor assemblies.

AFC's MC/IG Cable offers cost savings of up to 50% over pipe and wire and may be used in cable trays, above hung ceilings in accordance with NEC Section 300-22(c), and under computer room floors.

RED
Fire Alarm /Control Cable™

Designed for use in environmental air spaces, new Fire Alarm/Control Cable complies with Article 334 of the NEC as Type MC Cable and is in accordance with Section 518-4, of the NEC, for places of public assembly. This cable is now fully plenum rated, fire resistant, and low smoke.

The preferred replacement for pipe and wire in fire alarm runs, this new AFC Fire Alarm/Control Cable is listed and labeled by Underwriters Laboratories as Type MC Cable per Standard 1569 and Type FPLP per Standard 1424.

Jacketed MC Cable

Jacketed type MC may be used exposed or concealed, in cable trays or approved raceways, as an open run of cable, in wet, dry or oily locations, direct buried in earth or concrete or outdoor applications including messenger supported aerial installations. It is ideal for wet or corrosive atmosphere and perfect for site lighting, burial in slab construction and surface mounted or concrete buried in parking decks.

RED FIRE ALARM/ CONTROL CABLE™

UNIQUE RED ARMOR

Fire Alarm / Control Cable is now fully plenum rated, fire resistant and low smoke. It has passed the critical UL 910 Steiner tunnel test. One of the most critical components in any fire alarm system is run above suspended ceilings, within the walls or partitions, and under floors. This "component" is the electrical wiring system which interconnects pull stations, smoke detectors, alarm devices, bells, horns, and other equipment with the main fire alarm control panel.

Such all-important duty calls for the fire safety and performance of AFC's new metal-clad Fire Alarm/Control Cable which conforms to NEC Article 760, as detailed in Sections 760-14, -16, -18, -23, and -28.

Designed for use in environmental air spaces, new Fire Alarm/Control Cable complies with Article 334 of the NEC as Type MC Cable and is in accordance with Section 518-4, of the NEC, for places of public assembly.

The preferred replacement for pipe and wire in fire alarm runs, this new AFC Fire Alarm/Control Cable is listed and labeled by Underwriters Laboratories as Type MC Cable per Standard 1569.

Fire Alarm/Control Cable is competitive with Teflon® type cables and gives contractors a choice of sizes: #16, 18, 14, and 12 AWG (as permitted by NEC Section 430-72 for motors and Section 725-16 for other remote control circuits, and by 760-16 for fire alarm circuits) and larger if required. These cables are available in twisted pairs and twisted shielded pairs as well as conventional cabled conductors.

JACKETED MC CABLE

INTRODUCTION

For power, lighting control and signal circuits, AFC's jacketed MC cable, a single metal clad assembly, offers a cost effective replacement for pipe and wire. Designed for the maximum in circuit protection, identification and safety, jacketed metal clad cable with THHN/THWN insulated copper conductors provides: 1) a (90°C) dry, (75°C) wet, temperature rating, 2) an effective ground for sensitive applications through the internal copper grounding conductor, and 3) an overall PVC jacket to provide circuit identification and the maximum in physical protection.

DESCRIPTION

Copper conductors in sizes #14-10 AWG solid and #8-1/0 AWG stranded; THHN/THWN (90°C) insulation; bare or insulated ground wire; all conductors cabled with an internal marker tape; overall assembly tape; sheathed in a galvanized steel interlocked armor; and a PVC jacket overall (black PVC standard - other colors available upon request.)

Jacketed Type MC may be used exposed or concealed, in cable trays or approved raceways, as an open run of cable, in wet, dry or oily locations, direct buried in earth or concrete or outdoor applications including messenger supported aerial installations.